SEDIMENT FOR SURVIVAL

A Strategy for the Resilience of Bay Wetlands in the Lower San Francisco Estuary

I0063016

SFEI San Francisco Estuary Institute

PREPARED BY

Authors

SFEI
Scott Dusterhoff
Katie McKnight
Letitia Grenier

Strategic Adaptation Consultant
Nate Kauffman

Design and Cartography

SFEI
Ruth Askevold
Katie McKnight
Ellen Plane

FUNDED BY

San Francisco Bay Water Quality
Improvement Fund, EPA Region IX

A PRODUCT OF HEALTHY WATERSHEDS • RESILIENT BAYLANDS

APRIL 2021

SAN FRANCISCO ESTUARY INSTITUTE PUBLICATION #1015

SUGGESTED CITATION

Dusterhoff, S., McKnight, K., Grenier, L., and Kauffman, N. 2021. *Sediment for Survival: A Strategy for the Resilience of Bay Wetlands in the Lower San Francisco Estuary.* A SFEI Resilient Landscape Program. A product of the *Healthy Watersheds, Resilient Baylands* project, funded by the San Francisco Bay Water Quality Improvement Fund, EPA Region IX. Publication #1015, San Francisco Estuary Institute, Richmond, CA.

Version 1.3 (April 2021)

REPORT AVAILABILITY

Report is available at sfei.org/documents/sediment-for-survival

CONTENTS

Appendices are available at
www.sfei.org/documents/sediment-for-survival

Acknowledgements

This report is a product of the Healthy Watersheds, Resilient Baylands project, which is funded by the U.S. Environmental Protection Agency (EPA) Region IX's San Francisco Bay Water Quality Improvement Fund. Additional funding for this effort was provided by the Regional Monitoring Program for Water Quality in San Francisco Bay (RMP). We give special thanks to Luisa Valiela, the EPA project manager, for all of her support and enthusiasm throughout the project. We also thank Darcie Luce (San Francisco Estuary Partnership [SFEP]), the project contract manager, for all of her hard work on this effort and for being such a great project partner.

The Technical Advisory Committee (TAC) for this effort provided invaluable technical assistance and guidance through a series of meetings and provided helpful comments on draft materials. The TAC members were Josh Collins (SFEI), Maureen Downing-Kunz (United States Geological Survey [USGS]), Lorraine Flint (USGS), Barry Hecht (Balance Hydrologics), Noah Knowles (USGS), Jeremy Lowe (SFEI), Lester McKee (SFEI), Michelle Orr (Environmental Science Associates [ESA]), David Schoellhamer (USGS), and Karen Thorne (USGS). We also thank the Management Advisory Committee (MAC) for their important contributions. The MAC members were Donna Ball (SFEI

and South Bay Salt Pond Restoration Project [SBSPRP]), Jessica Davenport (State Coastal Conservancy [SCC]), Setenay Frucht (San Francisco Bay Regional Water Quality Control Board [SFBRWQCB]), Brenda Goeden (San Francisco Bay Conservation and Development Commission [BCDC]), Dave Halsing (SBSPRP), Roger Leventhal (Marin Dept of Public Works), Darcie Luce (SFEP), Brett Milligan (UC Davis), Heidi Nutters (SFEP), Sandra Scoggin (San Francisco Bay Joint Venture [SFBJV]), Renee Spenst (Ducks Unlimited), Christina Toms (SFBRWQCB), and Luisa Valiela (EPA). Additional project support was provided by John Bourgeois (ESA), Brenda Buxton (SCC), Susan De La Cruz (USGS), Naomi Feger (SFBRWQCB), Xavier Fernandez (SFBRWQCB), Amy Foxgrover (USGS), Matt Gerhart (SCC), Andy Gunther, and Bruce Jaffe (USGS). We owe a particular debt of gratitude to Lorraine Flint, Michelle Stern (USGS), and John Callaway (University of San Francisco) for providing unpublished data that were essential for our analyses.

We are also grateful to all the SFEI staff members who helped with data collection, mapping, analysis, reporting, document production, and project management. We especially thank Robin Grossinger, Steve Hagerty, Rusty Holleman, Jen Hunt, Pete Kauhanen, Sarah Pearce, Ellen Plane, Micha Salomon, Erica Spotswood, Phil Trowbridge, Alison Whipple, Taylor Winchell, and David Ludeke (SFEI intern from the Bill Lane Center for the American West at Stanford University).

Imagery of Corte Madera marsh, captured by drone. Imagery by Pete Kauhanen, SFEI.

EXECUTIVE SUMMARY
Sediment: A Crisis on the Horizon

The resilience of San Francisco Bay shore habitats, such as tidal marshes and mudflats, is essential to all who live in the Bay Area. Tidal marshes and tidal flats (also known as mudflats) are key components of the shore habitats, collectively called baylands, which protect billions of dollars of bay-front housing and infrastructure (including neighborhoods, business parks, highways, sewage treatment plants, and landfills). They purify the Bay's water, support endangered wildlife, nurture fisheries, and provide people access to nature within the urban environment. Bay Area residents showed their commitment to restoring these critical habitats when they voted for a property tax to pay for large-scale tidal marsh restoration. However, climate change poses a great threat, because there may not be enough natural sediment supply for tidal marshes and mudflats to gain elevation fast enough to keep pace with sea-level rise.

This report analyses current data and climate projections to determine how much natural sediment may be available for tidal marshes and mudflats and how much supplemental sediment may be needed under different future scenarios. These sediment supply and demand estimates are combined with scientific knowledge of natural physical and biological processes to offer a strategy for sediment delivery that will allow these wetlands to survive a changing climate and provide benefits to people and nature for many decades to come. The approach developed in this report may also be useful beyond San Francisco Bay because shoreline protection, flood risk-management, and looming sediment deficits are common issues facing coastal communities around the world.

Comparison of future bayland sediment demand, natural Bay sediment supply, and supply of additional sediment sources.

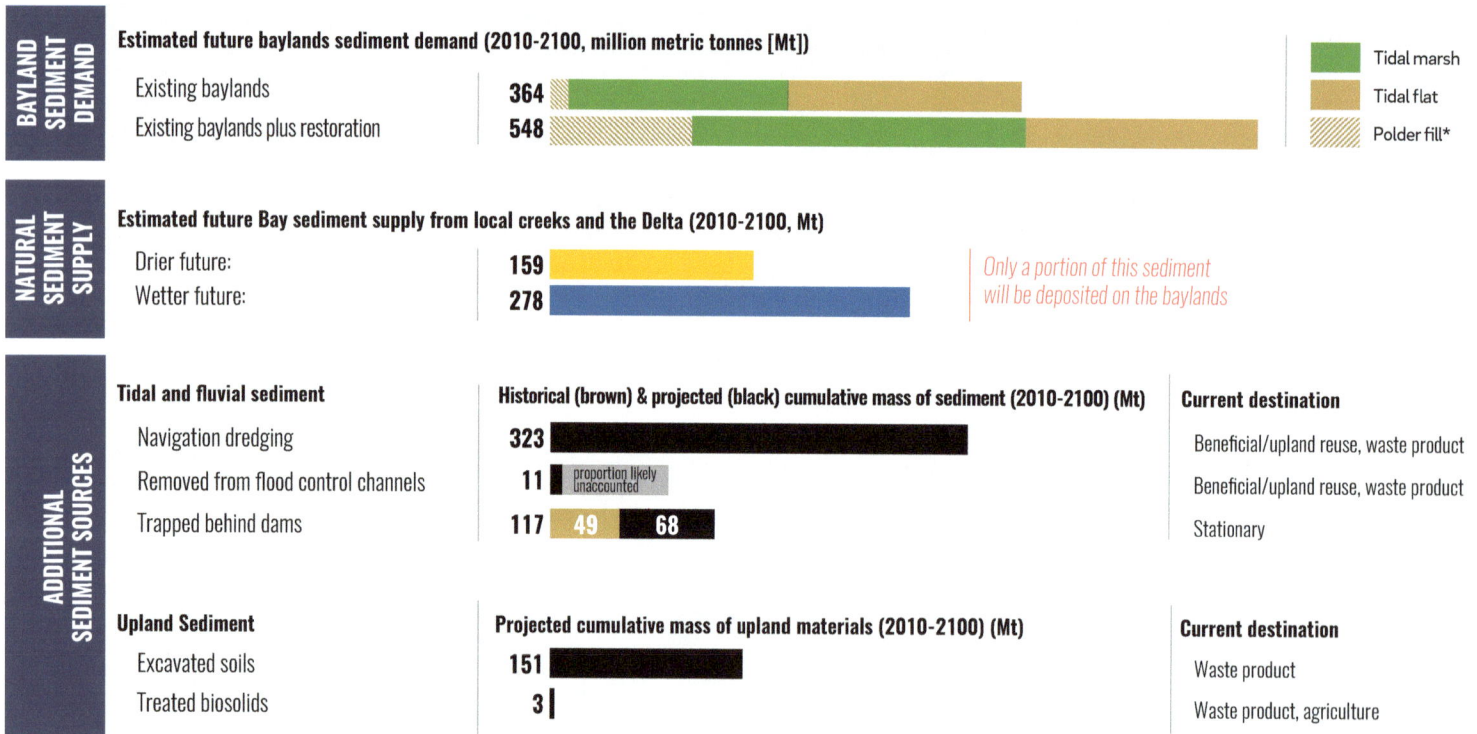

BAYLAND SEDIMENT DEMAND

Estimated future baylands sediment demand (2010-2100, million metric tonnes [Mt])

- Tidal marsh
- Tidal flat
- Polder fill*

| Existing baylands | 364 |
| Existing baylands plus restoration | 548 |

NATURAL SEDIMENT SUPPLY

Estimated future Bay sediment supply from local creeks and the Delta (2010-2100, Mt)

| Drier future: | 159 |
| Wetter future: | 278 |

Only a portion of this sediment will be deposited on the baylands

ADDITIONAL SEDIMENT SOURCES

Tidal and fluvial sediment — Historical (brown) & projected (black) cumulative mass of sediment (2010-2100) (Mt) — **Current destination**

Navigation dredging	323	Beneficial/upland reuse, waste product
Removed from flood control channels	11 (proportion likely unaccounted)	Beneficial/upland reuse, waste product
Trapped behind dams	117 / 49 / 68	Stationary

Upland Sediment — Projected cumulative mass of upland materials (2010-2100) (Mt) — **Current destination**

| Excavated soils | 151 | Waste product |
| Treated biosolids | 3 | Waste product, agriculture |

Polder fill is the sediment needed to bring deeply subsided areas (polders) slated for restoration up to tidal marsh elevation

Key Messages

- **Tidal marshes and mudflats are unlikely to receive enough sediment naturally to survive sea-level rise this century.** Restoring thousands of acres of historic marshes to the tides is invaluable for shoreline protection and the health of the Bay, but also increases overall sediment demand.

- **Other local sediment sources offer the potential to help maintain tidal marshes and tidal flats that will be resilient as the climate continues to change.** Preliminary analyses indicate that between now and 2100, sediment trapped in watersheds and dredged from the Bay, as well as soil excavated in construction projects around the region, will likely be greater than the amount of sediment arriving to the Bay from local rivers and the Delta.

- **Management practices need to change quickly to access these other sources of sediment that can help increase the future resilience of tidal marshes and mudflats.** Accessing these supplementary sediment sources will require rapid, unprecedented collaboration among public agencies, industry, and other stakeholders, as well as innovative approaches to sediment management and regulation. This report details a strategy for changing sediment management to increase the resilience of bay shore habitats and improve watershed health.

The range of sediment management actions that should be part of a multi-benefit sediment strategy.

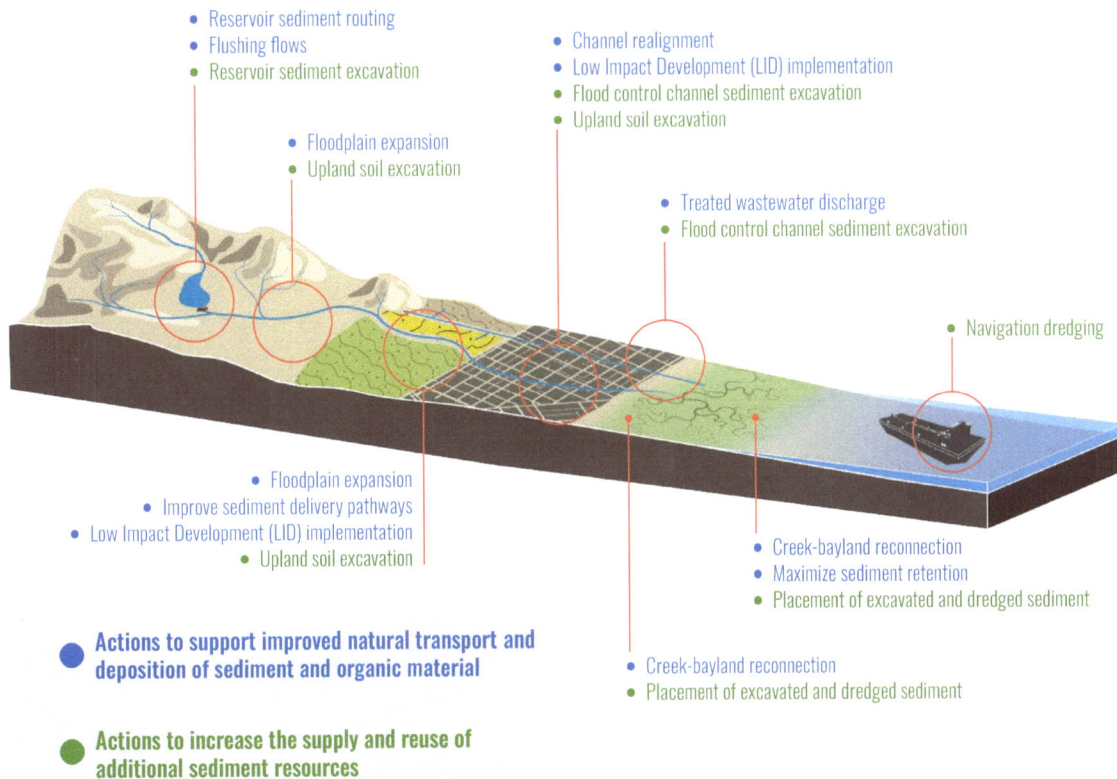

- Reservoir sediment routing
- Flushing flows
- Reservoir sediment excavation

- Channel realignment
- Low Impact Development (LID) implementation
- Flood control channel sediment excavation
- Upland soil excavation

- Floodplain expansion
- Upland soil excavation

- Treated wastewater discharge
- Flood control channel sediment excavation

- Navigation dredging

- Floodplain expansion
- Improve sediment delivery pathways
- Low Impact Development (LID) implementation
- Upland soil excavation

- Creek-bayland reconnection
- Maximize sediment retention
- Placement of excavated and dredged sediment

- Creek-bayland reconnection
- Placement of excavated and dredged sediment

● **Actions to support improved natural transport and deposition of sediment and organic material**

● **Actions to increase the supply and reuse of additional sediment resources**

Tidal flats at Palo Alto Baylands (Photo by Don DeBold, courtesy CC 2.0)

Introduction

Background

San Francisco Bay is home to thousands of acres of baylands that provide important fish and wildlife support, shoreline protection, water purification, and recreational opportunities. The baylands are a continuum of habitats that span the subtidal areas of the open bay and the intertidal areas of tidal flats and marshes. These habitats also buffer against storm surges and sea-level rise (SLR) by attenuating waves and protecting infrastructure located near the shoreline. As these habitat types that ring San Francisco Bay take on increasingly important roles, questions arise about whether or not they will persist under a changing climate and to what degree management will be necessary to conserve and expand them.

Numerous efforts have spurred widespread actions to restore and maintain healthy baylands, support endangered species recovery, and measure regional restoration progress. One of the earliest initiatives to establish a regional framework to improve the health of the Bay was the Comprehensive Conservation and Management Plan (CCMP) for the San Francisco Estuary (i.e., San Francisco Bay and Delta), an outcome of the federal Clean Water Act to identify actions needed to restore and maintain the integrity of the Estuary's biological, chemical, and physical processes (33 U.S.C. 1251, SFEP 1994, SFEP 2016). The CCMP resulted in the creation of the 1999 Baylands Ecosystem Habitat Goals project, the first region-wide effort to create goals and priorities for tidal wetland restoration in San Francisco Bay (Goals Project 1999). Through a collaborative vision for restoring 100,000 acres (~40,500 ha) of tidal marshes and other tidal habitats, the Goals Project sparked tens of thousands of acres of restoration in the region. In 2010, the San Francisco Bay Subtidal Habitat Goals Report expanded the work of the Goals Project by outlining a 50-year conservation plan for submerged areas of the Bay, a previously under-prioritized habitat zone (Subtidal Goals 2010). In 2015, the Goals Project was updated to incorporate new science-based recommendations to expand the focus from primarily marshes to include adjacent baylands habitats and to address climate change and other key physical and ecological drivers through the end of the century (Goals Project 2015).

Recently, regional efforts have expanded to improve cross-jurisdictional collaboration on nature-based SLR adaptation and promote multi-benefit actions that integrate habitat improvement and flood risk management at the Bay shore. Examples include the San Francisco Bay Conservation and Development Commission (BCDC) Strategic Plan Update, which has a goal of increasing the Bay's natural and built communities' resilience to rising sea level through a range of regionally coordinated efforts (BCDC 2017); the biannual State of the Estuary report, which tracks and updates metrics to assess overall health of the SF Bay-Delta Estuary (SOTER 2019); the Recovery Plan for Tidal Marsh Ecosystems of Northern and Central California, which outlines recovery strategies for five endangered flora and fauna species (USFWS 2013); Project Tracker and the Bay Area Aquatic Resource Inventory, which account for gains and losses in bayland habitat types due to permitted restoration, mitigation, conversion, and natural processes (CWMW 2020); and the San Francisco Bay Shoreline Adaptation Atlas report, a science-based framework for developing adaptation strategies that are appropriate for the diverse shoreline of the Bay and that take advantage of natural processes (SFEI and SPUR 2019). Flood Control 2.0, a multi-agency partnership to integrate habitat objectives within flood risk management actions, has also helped advance multi-benefit channel design and management through tools like SediMatch, numerous landscape visions, and regulatory and economic analyses (SFEI-ASC 2017a). The findings from DredgeFest California, a multi-stakeholder event organized by the Dredge Research Collaborative, highlight the critical need for sediment to support baylands and have informed recent bayland restoration project designs (Milligan et al. 2016). Additionally, strong regional and statewide political will to protect and preserve tidal habitats in San Francisco Bay has led to the creation of the San Francisco Bay Restoration Authority in 2008 and the approval by voters in 2016 of a 9-county parcel tax to fund shoreline projects that will protect, restore, and enhance San Francisco Bay. Although we have made great progress as a region in preserving and restoring our baylands, the threats that climate change poses to baylands are unprecedented, and the effects could be seen across a very short timescale.

Maintaining healthy baylands that survive into the future under a changing climate requires us to redesign our landscapes as robust, resilient systems that take advantage of natural processes to derive desired benefits. Supplying baylands with adequate sediment is perhaps the most critical element for their survival. Sediment has long been recognized as a precious and necessary resource for bayland survival (LTMS 1998, Goals Project 1999, Subtidal Goals 2010, Milligan et al. 2016, Public Sediment 2019), with many efforts to date focused on understanding the amount of sediment that will be needed in the coming decades. Knowles (2010) suggested that as much as approximately 13 Mcy/yr (10 Mm^3/yr) of organic matter and inorganic sediment may be needed to support existing tidal marshes around the Bay by 2100. The present deposition of inorganic sediment is approximately 0.26 Mcy/yr (0.2 Mm^3/yr) (Schoellhamer et al. 2005). In South Bay alone, Jaffe et al. (2011) projected approximately 0.92 Mcy/yr (0.7 Mm^3/yr) of sediment is needed to support existing baylands by 2100 and an additional 1.0 Mcy/yr (0.8 Mm^3/yr) is needed to support planned restoration. Perry et al. (2015) estimated that between 160 and 200 Mcy (122-153 Mm^3/yr) of sediment is needed just to bring 40,000 acres (~16,200 ha) of planned and in-progress tidal marsh restoration projects to present-day marsh plain elevation. The next step in bayland sediment science is developing an understanding of future bayland sediment need compared to future sediment supply for a range of climatic and management conditions. The impacts that changing precipitation patterns and SLR could have on the amount of sediment baylands need compared to sediment supply may require accelerated planning and coordination to allow San Francisco Bay's tidal habitats to persist through the end of the century.

Heron's Head during a King Tide, December 2014 (Photo by ebjSF, courtesy CC 2.0)

This report presents a Regional Sediment Strategy aimed at examining the future of mineral (inorganic) sediment in the Bay and informing sediment management for the resilience of tidal baylands (tidal marshes and tidal flats) to climate change and, in particular, SLR. This Regional Sediment Strategy is intended to complement other regional sediment science and strategy efforts and be a resource for bayland restoration managers, practitioners, planners, and regulatory agency staff. The Strategy focuses on addressing these questions:

1. What is the potential future bayland sediment need relative to future sediment supply throughout the Bay, and what is the potential impact of this ratio on bayland resilience?

2. What are appropriate sediment management approaches for delivering sediment to baylands to support resilience, and what are appropriate actions for promoting baylands resilience over time?

The amount of future sediment needed for baylands to accrete vertically and keep pace with SLR (sediment demand) and future Bay sediment supply is analyzed at different spatial and temporal scales for scenarios that consider climate change impacts (e.g., increasing sea level and changing precipitation and sediment supply) and planned marsh restoration projects. These findings are synthesized into an assessment of future bayland resilience with respect to vertical accretion, highlighting opportunity areas for high resilience under a range of future conditions. The report then provides a discussion of sediment management and monitoring considerations to help improve bayland resilience, including (1) actions to enhance sediment delivery to the baylands (ranging from promoting natural processes to more intensive active management); (2) key questions and knowledge gaps that need to be addressed in future research; and (3) regional efforts focused on monitoring bayland resilience. The report concludes with thoughts about the critical questions we as a region must address as we develop bayland management and restoration priorities.

This Regional Sediment Strategy is part of a larger EPA-funded effort for San Francisco Bay called *Healthy Watersheds, Resilient Baylands* (Figure 1.1). The *Healthy Watersheds, Resilient Baylands* project seeks to help reestablish landscape functions by working with nature to improve water quality, create habitat, provide flood protection to vulnerable communities, and reduce maintenance costs. The project offers an integrated approach to designing and implementing urban greening, wetland restoration, and water quality improvements within a set of representative urban watershed–bayland systems. This model of coordinated, multi-benefit projects is critical to achieving significant and lasting environmental outcomes in complex, interconnected systems with many jurisdictions and stakeholders. The most efficient way to invest in the innovations needed in our watersheds will be through coordinated expenditures by local flood management agencies, cities, private corporations, and others. This project takes another step towards leveraging those financial resources to accomplish demonstrable environmental and social benefits.

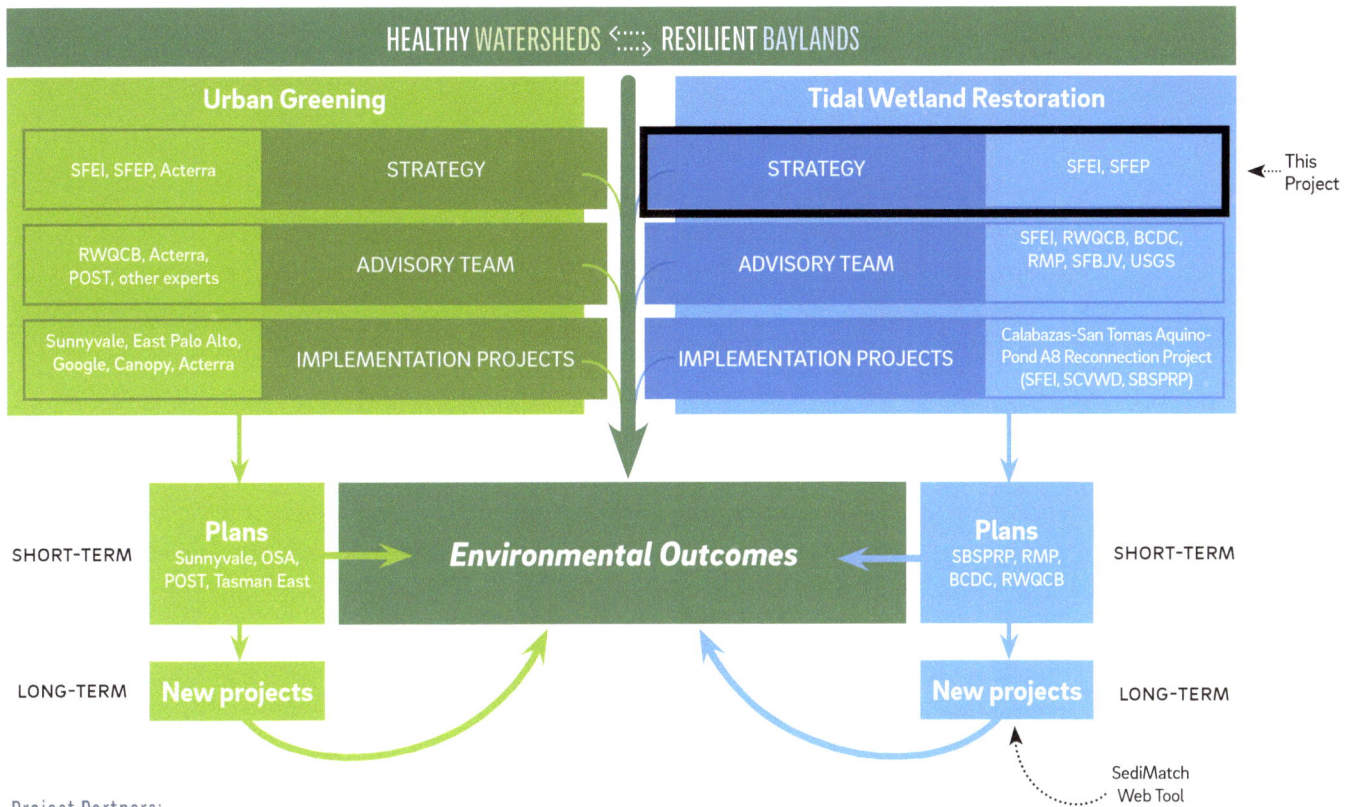

Figure 1.1. Structure of the Healthy Watersheds, Resilient Baylands project. For more information, visit sfei.org/projects/healthy-watersheds-resilient-baylands.

Project Partners:
Bay Conservation and Development Commission (BCDC)
San Francisco Bay Joint Venture (SFBJV)
San Francisco Estuary Institute (SFEI)
San Francisco Estuary Partnership (SFEP)
Santa Clara Valley Open Space Authority (OSA)
Santa Clara Valley Water District (SCVWD)
South Bay Salt Pond Restoration Project (SBSPRP)
Peninsula Open Space Trust (POST)
Regional Monitoring Program (RMP)
Regional Water Quality Control Board (RWQCB)
U.S. Geological Survey (USGS)

Healthy Watersheds, Resilient Baylands focuses on providing support for two categories of management actions: urban greening and tidal marsh restoration. The project includes development of a Multi-Benefit Urban Greening Strategy that incorporates ecological benefits into the Low Impact Development (LID) planning process and take advantage of a broader array of urban greening activities with hydrological benefits. It also includes the development of a Regional Sediment Strategy (this document) that synthesizes sediment need and availability data to maximize the value of limited sediment supplies in the design of tidal restoration projects for resilience to SLR. The two strategies have been applied to a series of innovative implementation projects along the South Bay shoreline and in the cities of Sunnyvale, Mountain View, and East Palo Alto. The project also includes further development of SediMatch, an innovative online marketplace for matching those who need sediment for bayland restoration projects with those who have sediment available (e.g., the dredging community). Once completed, the outputs from *Healthy Watersheds, Resilient Baylands* will be valuable tools for helping our local landscapes thrive in the future and provide benefits for both people and wildlife.

Environmental Setting

San Francisco Estuary, the largest estuary on the West Coast and one of the largest estuaries in North America, has a drainage area that includes almost half of the state of California (~60,000 mi^2 or ~155,000 km^2) (Conomos et al. 1985). The region experiences a Mediterranean climate (i.e., hot, dry summers and cool, mild winters) with strong regional precipitation gradients. In general, rainfall decreases from north to south and from west to east. North Bay watersheds experience considerably more rain than South Bay watersheds, and the western watersheds are more heavily influenced by the marine weather patterns than the eastern watersheds (Miles and Goudey 1997). The Bay currently receives approximately 2 million metric tonnes (Mt) of mineral sediment per year on average from the Delta and local tributaries, with tributaries supplying approximately two-thirds of the total and the largest contributors being Napa River, Sonoma Creek, Walnut Creek, and Alameda Creek (Schoellhamer et al. 2018) (Figure 1.2). Historically, however, the Delta supplied a much higher portion of the sediment load compared to local tributaries (Porterfield 1980). Hydraulic mining during the Gold Rush in the mid-1800s and other land modifications in the Sierra Nevada and Central Valley caused an increased sediment supply to the Bay during the mid-eighteenth to the late nineteenth centuries. The development of large dams and the success of erosion control measures substantially reduced the sediment supply to the Bay over the last two decades (Schoellhamer 2011). Over the past century, climate change has increased San Francisco Bay's mean tide elevation by more than 220 mm (~8.7 in) (Flick et al. 1999), while wintertime storm intensity has increased by 10–20% throughout much of the region (Russo et al. 2013).

Currently, there are approximately 80,000 acres (~32,400 ha) of baylands (i.e., tidal marshes and tidal flats) throughout San Francisco Bay (Figure 1.3). The majority of the baylands are dominated by tidal salt marshes, with salinity transitioning from saline to brackish to fresh from San Pablo Bay into Suisun Bay moving away from the Golden Gate and towards the Delta (Schile 2012). San Pablo Bay has the greatest extent of tidal marshes (both existing and recently restored areas returning to tidal marshes) while South Bay and Lower South Bay together have the greatest extent of tidal flats (SFEI-ASC 2017b). The low salinity marshes of Suisun Bay have a relatively high biomass content and therefore a relatively high local organic matter supply that helps maintain marsh elevations. The diking of baylands in the 19th and 20th centuries for agriculture, development, and commercial salt production resulted in an approximate 65% reduction in bayland extent compared to the early 19th century (Goals Project 1999). Many of these diked baylands are now deeply subsided areas known as polders, with the ground surface of some lying several feet below mean sea level. Polders around the Bay slated for restoration to tidal marsh will require a considerable amount of either naturally deposited or mechanically placed sediment just to get the ground surface at an appropriate tidal elevation for marsh plant establishment. To maintain suitable elevation into the future, marshes might require further additions of inorganic sediment to counter ongoing rapid SLR.

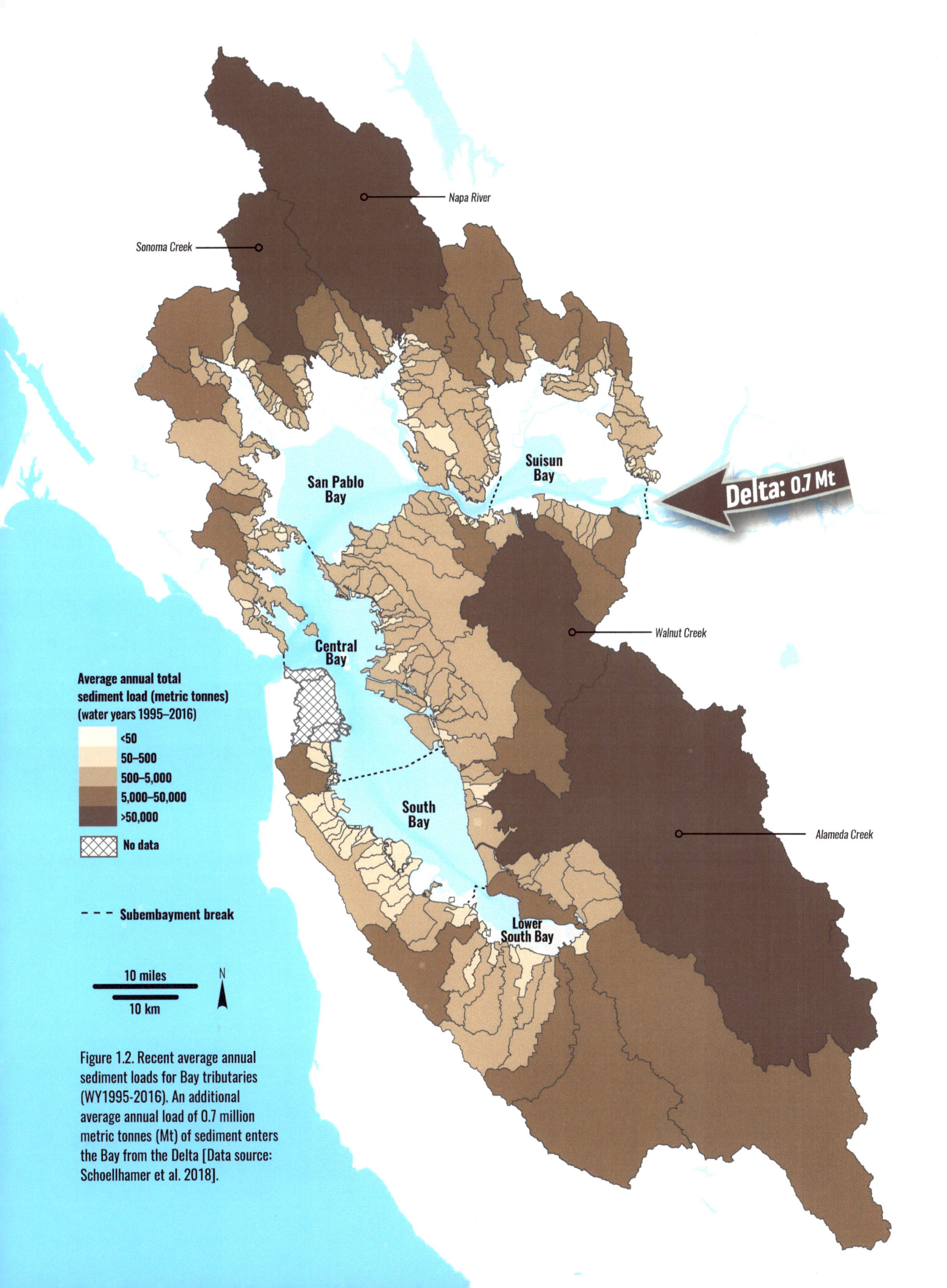

Napa River

Sonoma Creek

Suisun Bay

San Pablo Bay

Delta: 0.7 Mt

Central Bay

Walnut Creek

Average annual total sediment load (metric tonnes) (water years 1995–2016)

<50
50–500
500–5,000
5,000–50,000
>50,000

No data

South Bay

Alameda Creek

- - - **Subembayment break**

10 miles

10 km

N

Lower South Bay

Figure 1.2. Recent average annual sediment loads for Bay tributaries (WY1995-2016). An additional average annual load of 0.7 million metric tonnes (Mt) of sediment enters the Bay from the Delta [Data source: Schoellhamer et al. 2018].

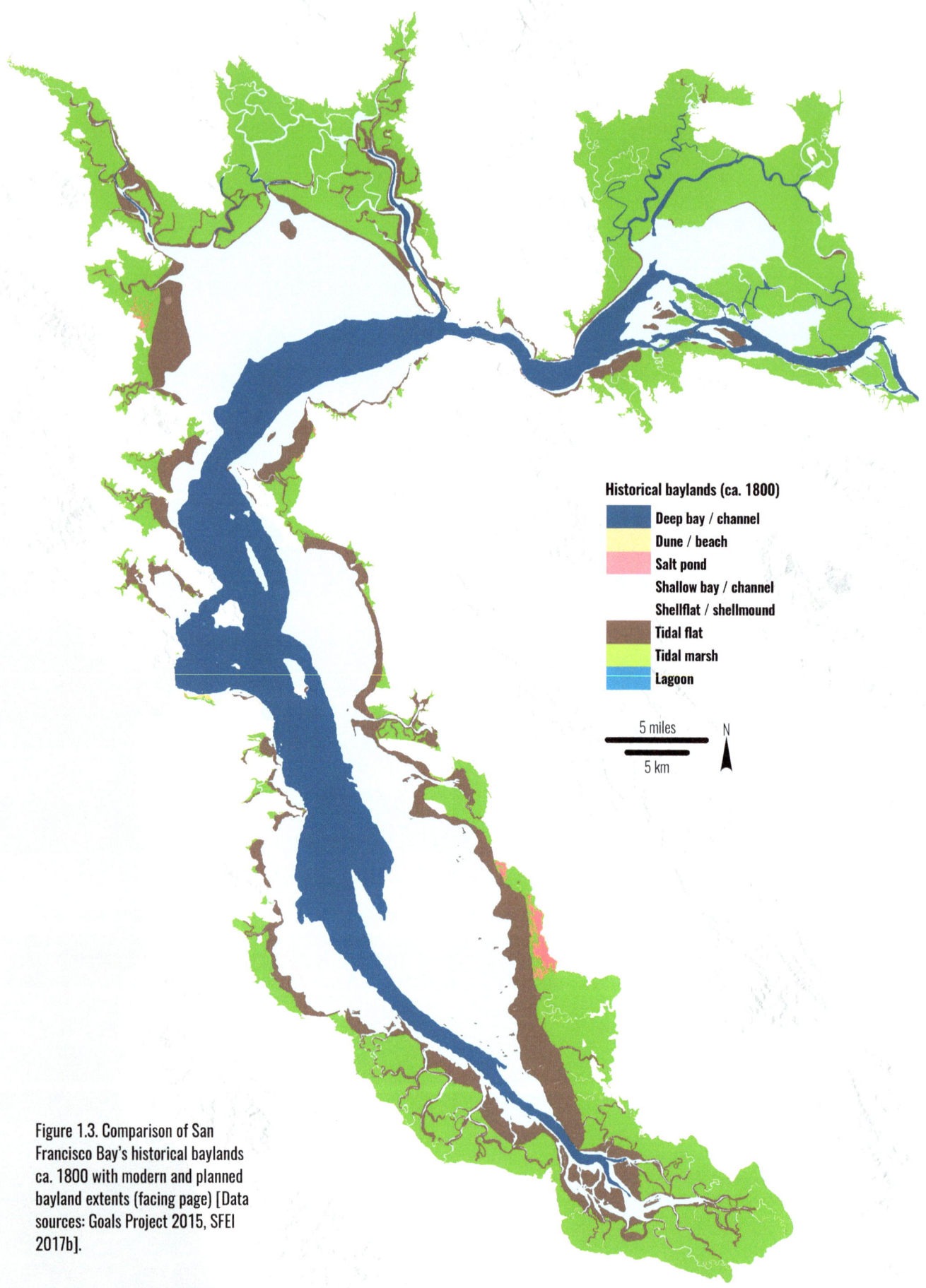

Historical baylands (ca. 1800)

- Deep bay / channel
- Dune / beach
- Salt pond
- Shallow bay / channel
- Shellflat / shellmound
- Tidal flat
- Tidal marsh
- Lagoon

5 miles

5 km

N

Figure 1.3. Comparison of San Francisco Bay's historical baylands ca. 1800 with modern and planned bayland extents (facing page) [Data sources: Goals Project 2015, SFEI 2017b].

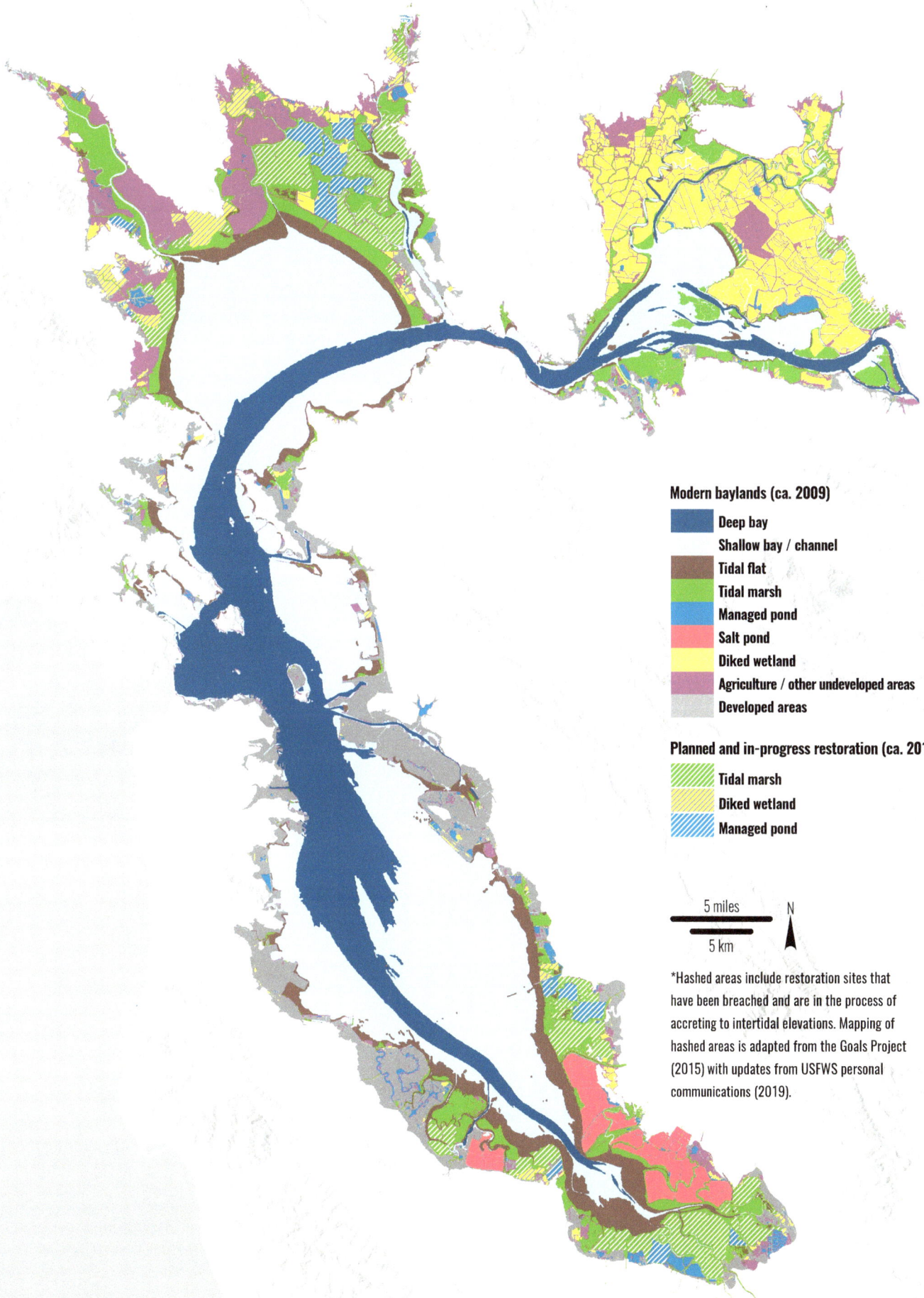

Modern baylands (ca. 2009)

- Deep bay
- Shallow bay / channel
- Tidal flat
- Tidal marsh
- Managed pond
- Salt pond
- Diked wetland
- Agriculture / other undeveloped areas
- Developed areas

Planned and in-progress restoration (ca. 2015)*

- Tidal marsh
- Diked wetland
- Managed pond

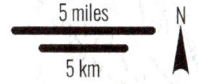

5 miles

5 km

N

*Hashed areas include restoration sites that have been breached and are in the process of accreting to intertidal elevations. Mapping of hashed areas is adapted from the Goals Project (2015) with updates from USFWS personal communications (2019).

Assessment of Bayland Resilience to Sea-Level Rise

Climate change and other landscape drivers will likely cause the baylands and their interconnected adjacent uplands to shift in location or convert to a different habitat type (Goals Project 2015), entering into a more dynamic period of landscape evolution compared to the 19th and 20th centuries. In order for the baylands to be resilient to SLR through the end of the century, they will need to (1) migrate landward; (2) prograde (i.e., expand into shallow subtidal areas of the Bay); and/or (3) accrete (i.e., gain tidal elevation, at least in-pace with SLR) (Brinson et al. 1995, Goals Project 2015) (Figure 1.4). These processes are not mutually exclusive and, in many places, they will be interconnected. However, the importance of each of these three processes will vary around the Bay. These processes are important to keep in mind as we conceptualize how baylands could evolve in the vertical, lateral, and upslope directions.

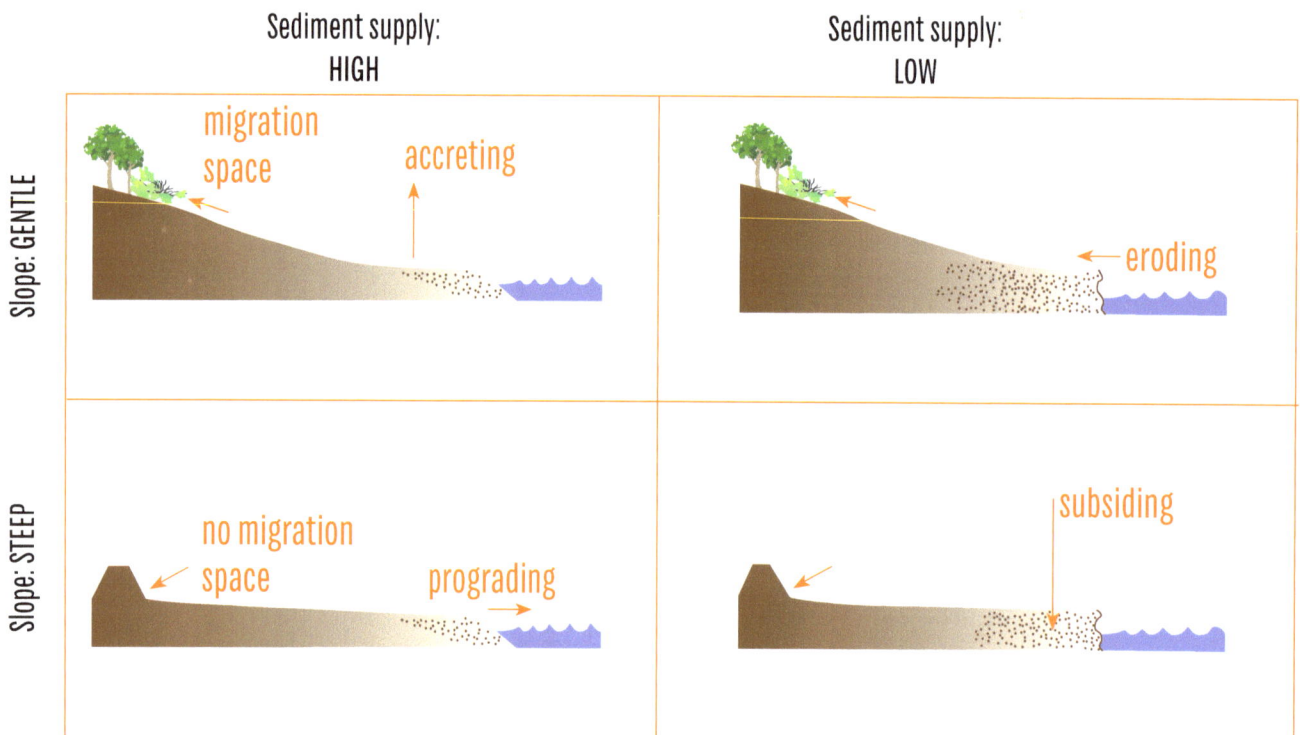

Figure 1.4. Conceptual diagram of the combination of ways baylands habitats could evolve under high and low sediment supplies with and without adequate space to migrate landward (adapted from Beagle et al. 2015).

Historical modification of the baylands has created major obstacles for marsh migration landward as sea level rises. Over 2,000 miles (~3,200 km) of engineered levees, berms, and other raised features exist along the Bay shore (SFEI 2016). Urbanization has led to many instances where development abuts tidal marsh, leaving little opportunity for landward migration (Orr et al. 2003, Stralberg et al. 2011, SFEI and SPUR 2019, Parker and Boyer 2017). Where undeveloped land borders tidal marsh, migration of wide marsh areas requires gentle topographic slopes that enable broad areas of colonization by tidal marsh vegetation keeping pace with rising water levels. SLR can "squeeze" marshes into narrow fringes where steep slopes or levees restrict their landward migration (Brinson et al. 1995, Stralberg et al. 2011, Goals Project 2015). In undeveloped locations with suitable topography, marsh migration can be challenged due to land ownership and development priorities (Goals Project 2015).

Many studies help explain the mechanisms for progradation and lateral erosion of the bayward margins of tidal marshes (e.g., Allen 1989, Schwimmer 2001, Bouma et al. 2016). However, measurements of these processes in the San Francisco Bay are limited to the south shore of Suisun Bay near Carquinez Strait, San Pablo Bay and parts of Central Bay within Marin County (Fischel and Robilliard 1991, Beagle et al. 2015, SFEI and Peter Baye 2020). The major factors affecting progradation or lateral erosion of the bayward edge of a marsh include wind-wave energy and direction, vegetation, sediment supply, tidal flat elevation, and SLR (Goals Project 2015, Lacy et al. 2020). While some studies hypothesize that lateral erosion (i.e., marsh retreat) is the main mechanism by which coastal wetlands are lost worldwide (Francalanci et al. 2013, Marani et al. 2011, Fagherazzi 2013), the evolution of lateral marsh movement with SLR in San Francisco Bay is understudied.

Vertical accretion of tidal marshes depends on many factors, including mineral sediment supply (Pestrong 1965, Krone 1987, Stralberg et al. 2011), organic matter accumulation (Patrick and DeLaune 1990, Callaway et al. 2012), inundation frequency (Duvall et al. 2019), vegetation dynamics (Pestrong 1965, Parker and Boyer 2017), soil compaction (Patrick and DeLaune 1990, Callaway et al. 2012), and land subsidence or uplift over time (Atwater 1977, Shirzaei and Burgmann 2018). With sufficient inorganic sediment and organic matter supply, shallow subtidal areas have naturally evolved into tidal marsh with vegetated plains at or slightly above local mean higher high water (MHHW), and can maintain these elevations for the range of SLR rates of the last two to three thousand years (Goals Project 1999, Byrne et al. 2001, Watson and Byrne 2009, Stralberg et al. 2011). Mineral sediments provide a source of nutrients to marsh vegetation, fueling root growth and organic matter production that, in turn, resists erosion and traps additional sediments (Patrick and DeLaune 1990). The depth, duration, and frequency of flooding is also an important factor for tidal marsh accretion, with higher rates of mineral sediment deposition closer to breach sites, tidal channels (Pestrong 1965; Goals Project 1999, 2015; Culberson et al 2004; Buffington et al. 2020), and creeks mouths (Temmerman et al. 2004). Compaction of underlying soils, which occurs naturally over time, can lead to the lowering of marsh surface elevations, with higher rates of consolidation observed in restored marshes compared to natural marshes (Patrick and DeLaune 1990, Orr et al. 2003). Elevational lowering can also occur due to local land subsidence from groundwater extraction or the presence of dikes, or due to regional tectonic shifts (Atwater 1977, Shirzaei and Burgmann 2018).

Several modeling efforts have evaluated whether tidal marshes in San Francisco Bay will be able to accrete vertically with SLR based on a wide range of scenarios. In a recent review of how Bay marshes will be impacted by SLR and climate change, Parker and Boyer (2017) note that two modeling efforts found suspended sediment concentrations and the rate of SLR to be key determinants of tidal marsh resilience (Stralberg et al. 2011, Swanson et al. 2015), with a third study (Schile et al. 2014) showing the importance of mineral sediment accumulation and plant productivity of organic matter in less saline parts of the Bay as a key determinant. The model results of the studies reviewed generally indicate Bay marshes high in the tidal frame will keep pace with low rates of SLR under moderate to high sediment supply by 2100, but under high rates of SLR, mid-high marsh habitats will shift to low marsh or migrate upslope while losing large extents due to limited migration space (Parker and Boyer 2019). While these models are helpful to understand the theoretical suspended sediment concentrations needed for marshes to keep pace under different rates of SLR, whether or not available mineral sediment will be deposited and remain on the marsh is challenging to discern.

For this effort, the assessment of bayland resilience addresses only the ability of tidal flats and tidal marshes to accrete vertically under a rising sea level. Our analyses focus primarily on the comparison of bayland sediment demand and Bay sediment supply, but we do include considerations of freshwater influence on organic matter accumulation and bayland accretion. Numerous studies have found that freshwater and brackish marshes have relatively high rates of organic matter accumulation which can help support marsh survival as sea level rises (Callaway et al. 1996, Orr et al. 2003, Stralberg et al. 2011, Callaway et al. 2012). Thus, we consider regional patterns in salinity gradients and marsh vegetation to inform a more detailed discussion on the relative resilience of bayland habitats to accrete vertically with SLR.

Marsh near Albany at low tide (Photo by tmikkphoto, courtesy CC 2.0)

Marsh near Albany at high tide (Photo by tmikkphoto, courtesy CC 2.0)

2

Los Alamitos Creek at Almaden Lake Park (Photo by Don DeBold, courtesy CC 2.0)

Bayland Sediment Demand and Bay Sediment Supply for Wetter and Drier Futures

The San Francisco Bay region has committed to restoring and maintaining tens of thousands of acres of tidal marsh over the next several decades (Goals Project 2015). Many of the areas slated for marsh restoration will require a large amount of sediment to bring them up to marsh elevation and sustain them as sea level continues to rise. This demand is in addition to the sediment needed for the tens of thousands of acres of existing tidal marshes and tidal flats to keep pace with sea-level rise. In this chapter, we examine possible futures of regional bayland sediment demand and sediment supply in order to understand the magnitude of potential bayland vulnerability as climate continues to change. Here we detail the methods and results for bayland sediment demand and supply analyses for the 21st century that consider sea-level rise, runoff and sediment delivery under a wetter and drier future, and bayland extent for existing conditions and a full restoration scenario.

Analysis Overview

Predicting the effects of climate change on bayland sediment demand and supply, and how these effects might in turn influence baylands resilience is complex and fraught with uncertainty. Depending on how and to what extent climate changes, bayland habitats could evolve in a number of ways. This study does not seek to determine what *will* happen with respect to future bayland sediment demand and sediment supply. Rather, it provides a range of bayland sediment demand and sediment supply estimates for different climatic conditions and bayland extents. The findings are based on the best available science and were developed with input and review by an interdisciplinary Technical Advisory Committee (TAC). The major assumptions, limitations, and uncertainties inherent to the analysis are presented in the Scenarios Analyzed section.

Estimates of sediment supply and transport tend to be reported in SI units of mass based on measurements of suspended sediment in the water column (e.g., Schoellhamer et al. 2018), whereas restoration practitioners focused on sediment demand and dredgers focused on sediment availability typically report their findings in English units of bulk volume (e.g., Perry et al. 2015, LTMS 2019, SediMatch 2020). In order to directly compare bayland sediment demand to supply, we report our findings in mass as opposed to volume.

Study Area

The study area includes the baylands of San Francisco Bay between the Golden Gate and the western boundary of the Sacramento-San Joaquin Delta at Broad Slough (Figure 2.1). The Delta east of Browns Island and Winter Island is excluded from this study. The baylands include all the intertidal environments, including tidal flats, beaches, and tidal marshes, plus the areas that would be intertidal if not for levees, dikes, seawalls, tide gates and other artificial structures that limit the landward excursion of the tides.

Under natural conditions, tidal flats extend from the local mean lower low water (MLLW) datum upslope to the bayward edge of intertidal vascular vegetation or non-vegetated beach (i.e., the foreshore), and include mudflats, sandflats, and shellflats. Tidal marshes extend from the landward edge of tidal flats or beaches to the maximum landward influence of the tides on species composition of the plant cover. The tidal elevation of the vegetated plain of an existing mature tidal marsh approximates the local MHHW datum, although natural levees along tidal marsh channels and marsh drainage divides are usually higher than MHHW. Diked baylands are formerly intertidal or shallow subtidal areas that were historically reclaimed but still exist at elevations below the local maximum tide height. Many diked baylands have subsided below their original tidal elevations.

Five major subembayments comprise San Francisco Bay: Suisun Bay, San Pablo Bay, Central San Francisco Bay (Central Bay), South San Francisco Bay (South Bay), and Lower South San Francisco Bay (Lower South Bay) (Figure 2.1). Breaks between subembayments used in this study were delineated based on existing Baylands Operational Landscape Unit (OLU) boundaries (as described in SFEI and SPUR 2019) in addition to suspended-sediment concentration monitoring site locations (as described in Schoellhamer et al. 2018).

Figure 2.1. The study area includes the watersheds and bayland habitats located within San Francisco Bay, from the Golden Gate to Broad Slough. Subembayments are based on Operational Landscape Unit (OLU) boundaries and USGS sediment flux monitoring locations.

Delta

Suisun Bay

San Pablo Bay

Central Bay

South Bay

Lower South Bay

PETALUMA

NAPA - SONOMA

SUISUN SLOUGH

MONTEZUMA SLOUGH

CARQUINEZ NORTH

NOVATO

GALLINAS

SAN RAFAEL

CORTE MADERA

PINOLE

CARQUINEZ SOUTH

Broad Slough

BAY POINT

WALNUT

WILDCAT

POINT RICHMOND

EAST BAY CRESCENT

RICHARDSON

Golden Gate

GOLDEN GATE

MISSION - ISLAIS

YOSEMITE - VISITACION

SAN LEANDRO

SAN LORENZO

ALAMEDA CREEK

COLMA - SAN BRUNO

SAN MATEO

MOWRY

BELMONT - REDWOOD

SAN FRANCISQUITO

STEVENS

SANTA CLARA VALLEY

Subembayment boundaries

OLU boundaries

OLU bayward boundaries

Planning for Sea-Level Rise Using an Operational Landscape Unit Approach

Planning for SLR often follows ownership or jurisdictional boundaries, but rising water levels will not necessarily follow these boundaries—and changes to the shoreline in one location may have unintended consequences in other locations. Instead, the scale of SLR planning should reflect the scale at which natural processes—such as tides, waves, and sediment transport—affect shorelines. An example of such an alternative planning scale is the Operational Landscape Unit (OLU) (Verhoeven et al. 2008).

Baylands OLUs, as defined by SFEI and SPUR (2019), share similar environmental variables—including topography, bathymetry, elevation, wave climate, shoreline characteristics, sediment supply, and adjacent land use—that influence their vulnerability and adaptability. OLUs often cross traditional jurisdictional boundaries of cities and counties, adhering instead to the boundaries of natural processes. Defining and delineating Baylands OLUs involved connecting watershed processes to the shoreline and into the Bay. Topography forms boundaries between watersheds, directing the flow of water and sediment in the uplands. However, in the marshes and mudflats of the baylands, the flatter topography and fine sediment processes tend to blur the boundaries between the Baylands OLUs. In some places the boundaries may be easily identifiable headlands, and in other places the boundary may be a fuzzy transition zone between adjacent creeks or tidal sloughs. Baylands OLUs consist of landscape features such as rivers, floodplains, and wetlands, as well as elements of the built environment such as parking lots, landfills, and residential neighborhoods (SFEI and SPUR 2019).

The connections between the features of the Baylands OLUs are important—altering the movement of sediment or water in one part of an OLU is likely to have an impact elsewhere in the OLU. For this reason, the OLU is a useful scale to plan for baylands resilience under changing freshwater and sediment supplies. This report quantifies potential future baylands sediment demand and future tributary sediment supply at the OLU scale to support coordination of restoration actions between projects within an OLU and regionally across the 30 OLUs delineated for San Francisco Bay. §

> **USEFUL RESOURCE:**
>
> *For more information on how OLUs were delineated for San Francisco Bay,* see Chapter 2 (page 21) of the San Francisco Bay Shoreline Adaptation Atlas *(SFEI and SPUR 2019).*

Scenarios Analyzed

Sea-Level Rise

The SLR projections used in this study are based on modeling for the San Francisco Bay Area by Kopp et al. (2014), and reflect the latest guidance adopted by the State of California (OPC 2018). We used 1.9 ft (~0.6 m) of SLR for the period 2010–2050, and 5.0 ft (~1.5 m) of SLR for the period 2050–2100. These projections reflect a high greenhouse gas emission scenario (Representative Concentration Pathway [RCP] 8.5). The probability of SLR meeting or exceeding these estimates by their corresponding time periods is low, approximately 0.5%, and they therefore support a high degree of precautionary shoreline planning to account for SLR (OPC 2018). Because projections are based on tide gage data for San Francisco Bay, regional rates of subsidence are included in the SLR estimates used (OPC 2018). These projections do not, however, take into account extreme SLR (i.e., the H++ scenario developed by Sweet et al. 2017).

Corte Madera Marsh (Drone imagery by Pete Kauhanen, SFEI)

Sediment Supply

This study assesses sediment supply in the 21st century for a wetter future and a drier future. The wetter future reflects the conditions in the CESM1-BGC RCP 8.5 downscaled global circulation model (GCM) and the drier future reflects the conditions in the HadGEM2-CC RCP 8.5 downscaled GCM. These GCMs were a focus of California's Fourth Climate Change Assessment (Pierce et al. 2018) and were selected for use in this study because they show moderate changes to future precipitation and runoff compared to the other models. Future Delta sediment loads for the two futures were derived from previous research by the USGS CASCaDE project (Stern et al. 2020). The future sediment loads from Bay tributaries were derived from a combination of modeled runoff for the two futures and sediment rating curves (i.e., relationships between runoff and sediment load).

Baylands Extents

This study quantifies future sediment demands for the baylands by considering bookend scenarios for bayland habitat extents. These are (1) existing tidal baylands plus those where restoration is in-progress; and (2) existing tidal baylands plus all diked baylands that have been purchased and are slated for restoration as of 2015 (i.e., as identified in the 2015 Baylands Goals Science Update). This report disregards diked baylands that, as of 2015, are not slated for restoration to tidal baylands. Thus, the calculations of future sediment supply and demand assume that these other baylands that will remain diked require minimal amounts of sediment. Bayland extents are adapted from the Bay Area Aquatic Resource Inventory (BAARI, Version 2.1), a regional dataset of tidal and non-tidal aquatic systems and riparian functional areas in San Francisco Bay (SFEI 2017b). More specifically, the extents of existing tidal flats and tidal marshes adapted from BAARI are largely based on interpretation of aerial imagery from 2009 and include some restoration updates that were in progress circa 2009. The inclusion of tidal flats is a first step towards a more detailed accounting of sediment needs across the "complete tidal marsh system" (Goals Project 2015), thereby recognizing the important role tidal flats play in supporting tidal marshes as well as the interconnectivity of these habitats. For example, tidal flats attenuate wind-generated waves and boat wakes that can erode marsh foreshores. Tidal flats also temporarily store sediment to be resuspended and transported into marshes during flood tides, and tidal marshes nurture tidal flats with organic matter and nutrients. Many species of fishes and birds move between tidal flats and tidal marshes as the tides rise and fall.

General Assumptions, Limitations, and Uncertainties

- This document is intended to support management decisions and does not constitute a management plan or vision.

- Due to the uncertainties in the variables that inform this analysis and the data gaps that exist, comparisons between sediment supply and demand should not exceed an order-of-magnitude granularity.

- Sediment supply and demand estimates are limited to mineral sediment (referred to henceforth as simply "sediment") and do not explicitly include estimates of organic matter.

- Analyses of future ratios between sediment supply and demand presented in Chapter 3 do not consider the amount of sediment needed to maintain managed (i.e., diked) wetlands.

- For computational purposes, a resilient tidal marsh is defined as an intertidal area that maintains elevations suitable for colonization by native vascular vegetation through the end of this century. A resilient tidal flat is as an intertidal area at lower elevations than tidal marsh that lacks vascular vegetation and persists in size and plan-form through this century. §

Sediment at Don Edwards San Francisco Bay National Wildlife Refuge (Courtesy of CC 2.0, photo by Allie Caulfied)

Bayland Sediment Demand Analysis
Methods and Assumptions
Habitat Scenarios

According to the latest available regional mapping (i.e., Bay Area Aquatic Resources Inventory), the study area has approximately 28,000 acres (~11,300 ha) of tidal flats, 45,000 acres (~18,200 ha) of tidal marsh, and 6,000 acres (~2,400 ha) of active restoration sites in the process of accreting to tidal marsh elevations (SFEI-ASC 2017b). If currently planned restoration projects outlined in the Goals Project (2015) are completed, they will add approximately 24,000 acres (~9,700 ha) of tidal marsh to previously mapped baylands habitats (Goals Project 2015), totaling approximately 75,000 acres (~30,400 ha) of tidal marsh and 28,000 acres (~11,300 ha) of tidal flats in the study area. While this does not achieve the aspirational goal of 100,000 acres (~40,500 ha) of tidal marsh outlined in the Goals Project (2015), it reflects the latest spatially-explicit regional data on planned restoration available. These estimates form the basis of the two baylands extent scenarios (i.e., alternative target extents for year 2100) used within our sediment demand analysis (Figure 2.2). The difference between the two scenarios is 24,000 acres (~9,700 ha) of tidal marsh:

- **Existing baylands:** 28,000 acres (~11,300 ha) of tidal flat, 51,000 acres (~20,600 ha) of tidal marsh (existing and evolving);

- **Existing baylands + planned restoration:** 28,000 acres (~11,300 ha) of tidal flat, 75,000 acres (~30,400) of tidal marsh.

Mass of Sediment for Each Habitat Scenario

A simple volumetric approach was used to quantify the future potential magnitude of baylands sediment demand for each habitat scenario, summarized in three main steps:

Step 1: Calculate the volume of soil needed to raise low-lying areas to restoration elevations. For this step, we computed the difference between existing and desired (i.e., restored) elevations and multiplied by area to determine the amount of polder fill needed to raise all active and planned restoration sites to appropriate intertidal elevations (Equation 1). Existing tidal flats and marshes, as determined by 2009 habitat data (SFEI 2017b) were excluded from this portion of the analysis because we assumed these habitats already exist at their equilibrium elevations within the tidal frame. We used present-day, local MHHW levels to set the desired elevations, which follows the recommendations of the Baylands Goals report (Goals Project 2015) to maximize elevation capital as much as possible to give marshes a better chance at maintaining their elevation over time. However, we recognize that marshes might be resilient at lower elevations. Elevation data used to determine existing conditions come from the Coastal National Elevation Database (CoNED) topobathymetric model of San Francisco Bay (USGS 2013), which largely reflects 2010 topobathymetric conditions. Additional bathymetric data of varying sources and resolutions was appended to this dataset to correct areas that likely reflect water surface elevations rather than true bathymetry (see Appendix A for more information). Polder fill volumes were converted to mass of mineral sediment by using bulk density estimates explained in Step 3 below.

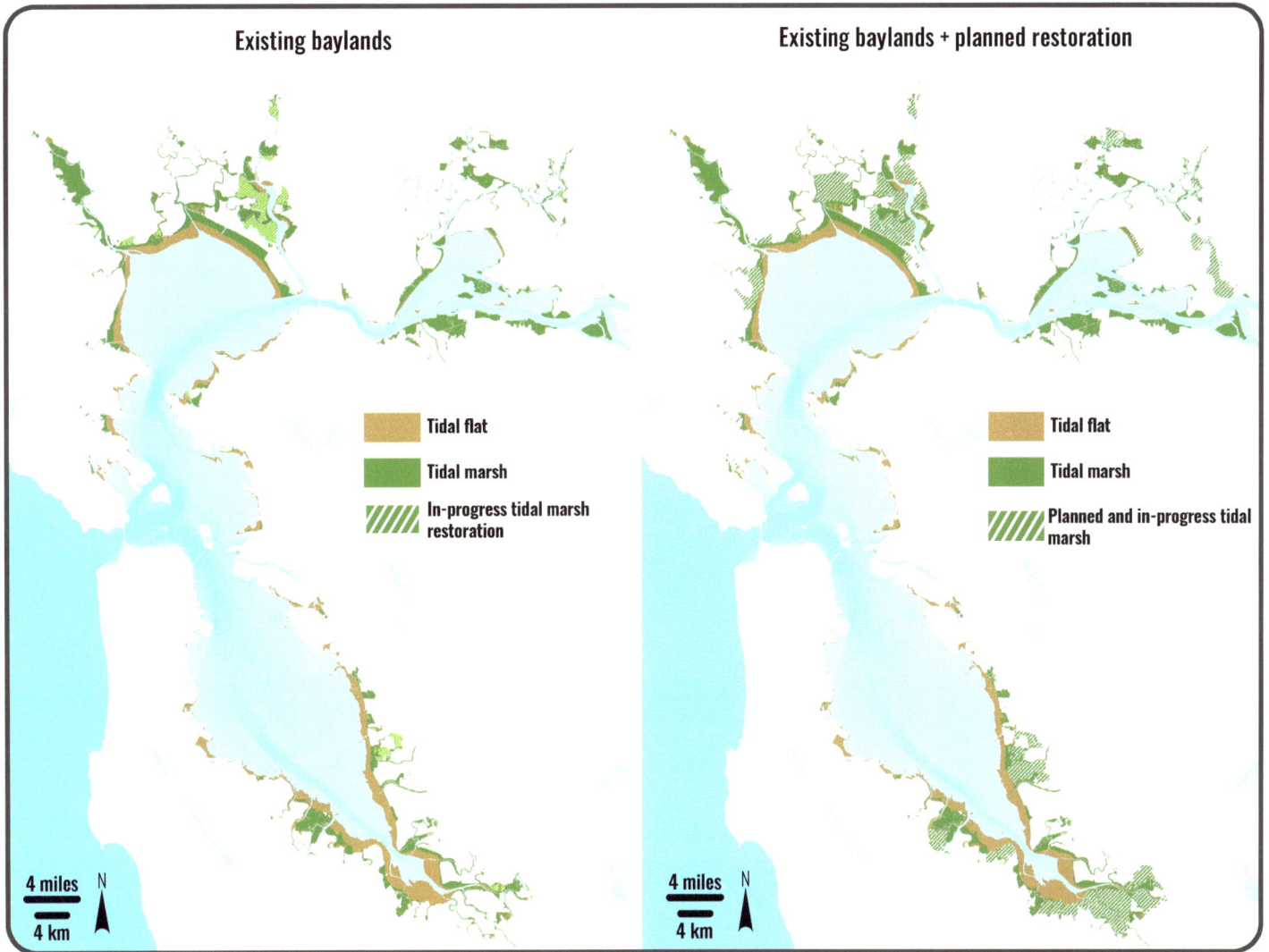

Existing baylands

Existing baylands + planned restoration

Legend (left map):
- Tidal flat
- Tidal marsh
- In-progress tidal marsh restoration

Legend (right map):
- Tidal flat
- Tidal marsh
- Planned and in-progress tidal marsh

4 miles / 4 km · N

Figure 2.2. Two baylands habitat scenarios were considered when quantifying future sediment demands: existing baylands (left) and existing baylands + planned restoration (right). The latter scenario includes the additional sediment needed for all lands purchased and slated for restoration as of 2015 (Goals Project 2015). Both scenarios consider the amount of sediment needed to raise low-lying areas to current tidal marsh elevations to achieve in-progress and planned restoration goals, indicated by hatching on the maps.

$$V_1 = (E_D - E_C) * A$$

V_1 = *volume of soil needed to raise active or planned restoration sites to desired elevations [m³]*

E_D = *desired, post-restoration elevation (i.e. local MHHW) [m]*

E_C = *existing, pre-restoration elevation [m]*

A = *habitat area [m²]*

Equation 1. Volume of soil needed to raise active or planned restoration sites to equilibrium elevations.

Step 2: Calculate the volume of soil needed for baylands to maintain their position within the tidal frame. After calculating the amounts of polder fill needed to raise active and planned restoration sites to desired elevations, we calculated the volume of soil needed for baylands habitats to maintain their position within the tidal frame over time (Equation 2). To do this, we multiplied the habitat acreages for both bayland extents (i.e., existing baylands, and existing baylands + planned restoration) by SLR projections of 0.6 m (~1.9 ft) for the period 2020–2050, and an additional 1.5 m (~5.0 ft) for the period 2050–2100, totaling 2.1 m (~6.9 ft) of SLR between now and the end of century. This was used to derive the volume of sediment needed for tidal marsh to keep pace with SLR in the near term (2010–2050) and the long term (2050–2100) (Figure 2.3).

$$V_2 = A(r) + V_1$$

V_2 = volume of soil needed to keep pace with SLR [m³]
A = habitat area [m²]
r = height of projected SLR [m]
V_1 = volume of soil needed to raise active or planned restoration sites to desired elevations [m³]

Equation 2. Volume of soil needed to keep pace with SLR.

$$b = (1-p)*d$$

b = average mineral sediment component of dry bulk density [t/m³]
p = percent organic matter content [%]
d = dry bulk density [t/m³]

Equation 3. Calculation of the average mineral sediment component of dry bulk density.

$$M = V_2(b)$$

M = mass of mineral sediment [t]
V_2 = volume of soil needed to keep pace with SLR [m³]
b = average mineral sediment component of dry bulk density [t/m³]

Equation 4. Conversion from volume of soil to mass of mineral sediment.

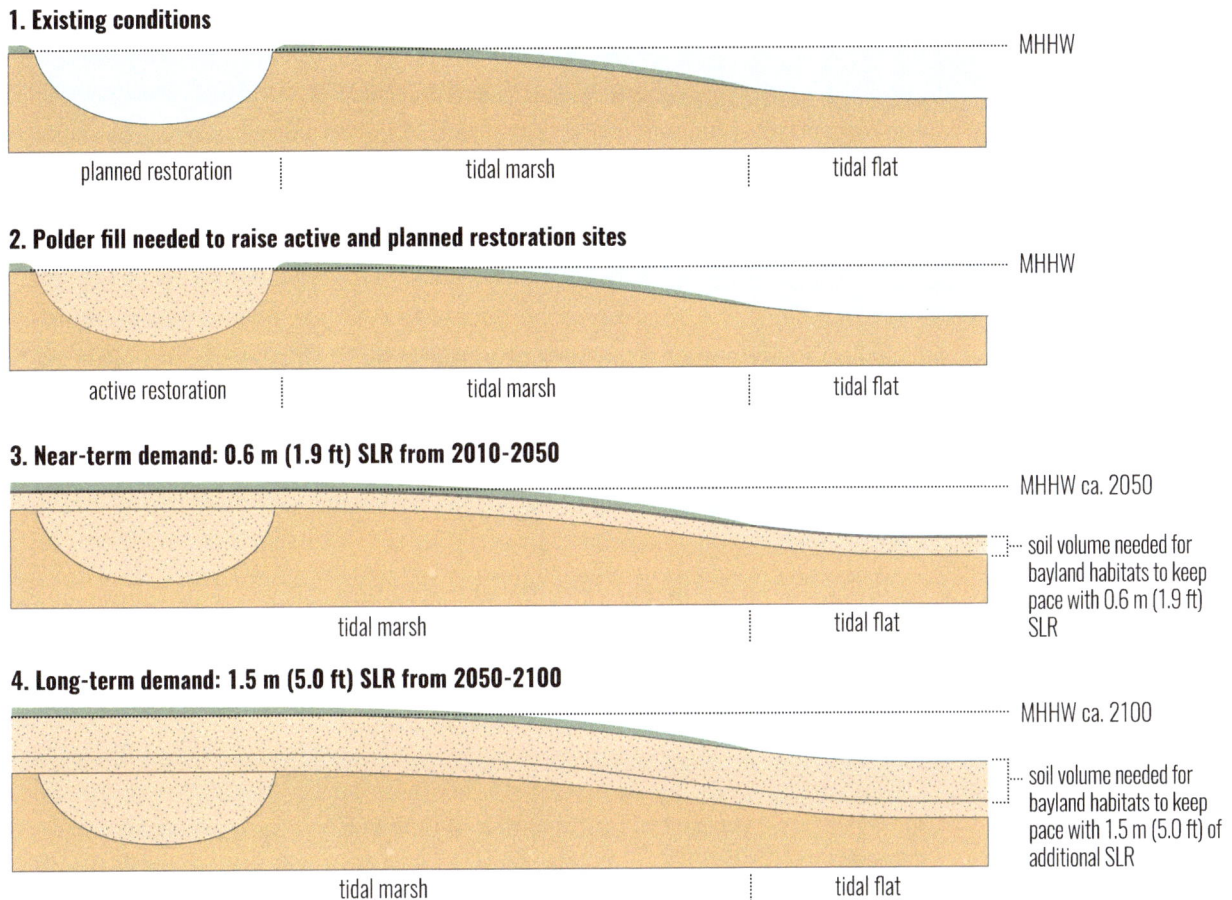

1. Existing conditions

planned restoration | tidal marsh | tidal flat

MHHW

2. Polder fill needed to raise active and planned restoration sites

active restoration | tidal marsh | tidal flat

MHHW

3. Near-term demand: 0.6 m (1.9 ft) SLR from 2010-2050

tidal marsh | tidal flat

MHHW ca. 2050

soil volume needed for bayland habitats to keep pace with 0.6 m (1.9 ft) SLR

4. Long-term demand: 1.5 m (5.0 ft) SLR from 2050-2100

tidal marsh | tidal flat

MHHW ca. 2100

soil volume needed for bayland habitats to keep pace with 1.5 m (5.0 ft) of additional SLR

Figure 2.3. Conceptual sections depicting soil volumes needed at different steps in this analysis. The top two sections depict the amount of polder fill needed to raise a planned restoration site to local MHHW levels. The bottom two sections depict the volume of soil needed for existing tidal flats and marshes (in addition to newly restored sites) to maintain their position in the tidal frame with near- and long-term SLR projections (Note: sections not to scale).

Step 3: Convert volumes of soil to mass of mineral sediment using local bulk density estimates for each habitat type. To allow for comparisons between bayland sediment demand and regional sediment supply, we converted total soil volumes calculated in steps 1 and 2 to mass of the inorganic sediment component using dry bulk density estimates averaged by subembayment for each habitat type (Equations 3 and 4). Bulk density estimates are based on data from Lionberger and Schoellhamer (2009) for tidal flats and data from Callaway et al. (2012) for tidal marshes, as described in Table 2.1 below.

For tidal marsh habitats, bulk density estimates varied within each subembayment for the near- and long-term time horizons. In the near term (0.6 m (~1.9 ft) of SLR), bulk densities were averaged across the top 0.2 m (~0.7 ft) of soil cores sampled by Callaway et al. (2012) (Table 2.1). In the long term (2.1 m (~6.9 ft) of SLR), bulk density estimates were averaged across a longer depth, the top ~0.4 m (~1.4 ft) of soil cores (i.e., the deepest depth in common across all cores sampled by Callaway et al. (2012)) in order to account for some degree of surface sediment compaction over time (unpublished soil core data, John Callaway, pers. comm.). Surface sediment compaction was not taken into account for tidal flat bulk density estimates since the amount of soil organic matter found in tidal flats is assumed to be minimal, making compaction less of a concern within tidal flat habitats. Therefore, tidal flat bulk density estimates remained the same between near- and long-term time horizons for each subembayment. More details on these methods are available in Appendix A.

We summarize bayland sediment demand by reporting the mass of mineral sediment needed for tidal flats and marshes in each subembayment to reach and sustain marsh plains at MHHW. We report estimates at the Bay scale and the subembayment scale to evaluate how future demand may vary across geographies. We also quantify sediment demand by OLU, which is the scale that we use to compare local tributary sediment supplies to local bayland sediment demands in the next chapter.

Table 2.1. Bulk density estimates used to convert soil volumes to mass of mineral sediment for each habitat type averaged by subembayment. Tidal marsh bulk density estimates are adapted from Callaway et al. (2012) and reflect the mineral component of dry bulk density. Tidal flat bulk density estimates are adapted from Lionberger and Schoellhamer (2009) and reflect bed density estimates that include organic matter and are based on measurements from Caffrey (1995), Sternberg et al. (1986) and Bruce Jaffe (pers. comm.). For more information on methods, see Appendix A.

Subembayment	Average mineral sediment component of dry bulk density of tidal marsh: near-term estimates based on top 0.20 m (0.7 ft) of soil core data [t of sediment/m³ of soil] (lbs/ft³)	Average mineral sediment component of dry bulk density of tidal marsh: long-term estimates based on top 0.44 m (1.4 ft) of soil core data [t of sediment/m³ of soil] (lbs/ft³)	Average sediment bed density of tidal flats [t of sediment/m³ of soil] (lbs/ft³)
Suisun Bay	0.16 (10.2)	0.23 (14.4)	0.86 (53.9)
San Pablo Bay	0.40 (25.2)	0.42 (26.4)	0.73 (45.6)
Central Bay	0.46 (28.8)	0.46* (28.8)	1.00 (62.2)
South Bay	0.47 (29.4)	0.51 (31.5)	0.77 (48.3)
Lower South Bay	0.47** (29.4)	0.51** (31.5)	0.58 (36.3)

*Bulk density estimate decreased from 0.46 to 0.38 t/m³ when averaged across 0.44 m (1.4 ft) of soil cores compared to the top 0.20 m (0.7 ft) of soil cores collected within Central Bay. Because bulk density is expected to increase with compaction, the near-term bulk density estimate for Central Bay was used to calculate sediment demand for the long-term scenario.

**No marsh samples in Lower South Bay were reported by Callaway et al. (2012) so bulk density values were adapted from the South Bay marshes for soil volume conversions in the tidal marsh habitats in Lower South Bay.

What is bulk density?

(excerpt from McKnight et al. 2020)

Soil is comprised of three main components: mineral sediment (sand, silt, clay, shell fragments), organic material (e.g., plant detritus, plant roots), and pore space, which can be filled with air or water (Cohen 2008) (Figure 2.4). Bulk density is a measure of the mass of total mineral sediment and organic material within a defined volume of soil. Differences exist between estimates of bulk density depending on how it is measured. Fully saturated bulk density and dry bulk density are estimates of both mineral sediment and organic matter. The mineral component of dry bulk density is an estimate of only the mineral sediment present in a given soil volume, thus all of the organic matter has been removed.

Bulk density is key in converting from soil volumes to mass of mineral sediment. Compaction, pore size, organic matter accumulation, and water moisture are some of the factors that affect the bulk density of soil and cause it to change over space and time. Bulk density can vary across habitat types with changes in vegetation and it can vary with changes in elevation and inundation patterns. Bulk density typically increases with depth and over time with compaction (McKnight et al. 2020). §

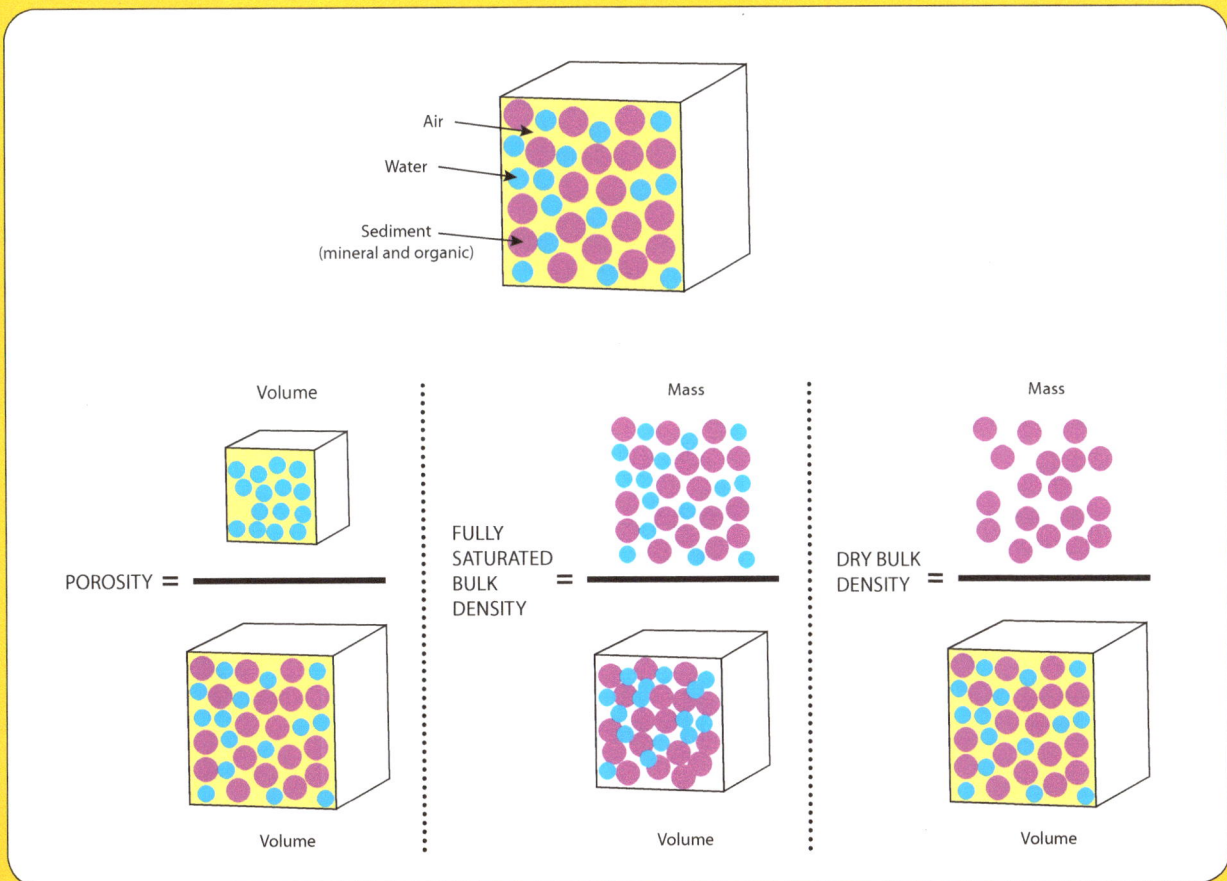

Figure 2.4. (Top) Conceptual diagram of the main components of soil. The sediment depicted in purple is made up of both organic matter and mineral sediment. (Bottom) Porosity, full saturated bulk density, and dry bulk density are soil measurements that can be used to estimate the mass of sediment in a given volume of soil. The equations for each measurement are depicted above (courtesy of McKnight et al. 2020; adapted from Flemming and Delafontaine 2016).

FUTURE SEDIMENT DEMAND:
Key Assumptions and Considerations

- Sediment demand estimates are limited to tidal flat and tidal marsh habitats and do not include the amount of sediment needed for adjacent connected habitats to keep pace as sea level rises. The volumetric approach used does not capture dynamic processes or feedback loops.

- The data used to delineate habitat acreages for the existing tidal flat and tidal marsh habitats and in-progress restoration (i.e., sites that have been breached and are in the process of accreting to tidal marsh elevations) are based on the Bay Area Aquatic Resources Inventory dataset (Version 2.1) which largely reflects 2009 habitat conditions (SFEI 2017b).

- All planned tidal marsh restoration is limited to permitted or acquired sites as of 2015, as specified in the Baylands Goals Update (Goals Project 2015) and does not include planned restoration beyond 2015 (e.g., Sonoma Creek baylands restorations).

- Sediment demand estimates to raise planned tidal marsh restorations are based on topobathymetric data from 2010 and thus do not account for the sediment placed at restoration sites over the past decade. In particular, the ~1.3, 0.8, and 4.5 Mcy (~1.0, 0.6, and 3.4 Mm^3) of sediment placed at Hamilton, Cullinan Ranch, and Montezuma restoration projects respectively between 2010 and 2017 are not included in demand estimates. Additionally, polder fill estimates will only grow with time, as MHHW levels increase with SLR. There is a benefit to restoring these sites sooner rather than later.

- The sediment demand calculations assume all existing tidal flat and tidal marsh habitats will maintain their habitat types through the end of the century, and all in-progress or planned tidal marshes will reach and maintain local MHHW elevations by 2050 through the end of century and at pace with SLR.

- Sediment demand calculations for the long-term time period (2050–2100) assume that all baylands in 2050 will be at elevations suitable for tidal flats and tidal marshes, as mapped in Figure 2.2.

- In-situ bulk density data for intertidal habitats in San Francisco Bay are limited and thus considerable uncertainty is introduced when converting soil volumes to mass of mineral sediment. §

Results and Discussion

Future Baylands Sediment Demand at the Bay Scale

Sediment needed for existing baylands to keep pace with SLR projections: Approximately 364 Mt of sediment are needed for the 28,000 acres (~11,000 ha) of tidal flats and 51,000 acres (~21,000 ha) of tidal marshes of existing bayland habitats to keep pace with 2.1 m (6.9 ft) of SLR between 2010 and 2100 (Figure 2.5). About 30% of this estimate—approximately 106 Mt—is needed in the near term (by 2050), and the remaining 70%—approximately 257 Mt—is needed in the long term (by 2100). Sediment needed to fill polders in existing, in-progress restoration sites to tidal marsh elevations accounts for approximately 14 Mt of the sediment demand, comprising around 4% of the overall demand throughout both time periods in this restoration scenario. This demand largely reflects the sediment needed to raise currently breached portions of the Eden Landing and Napa Ponds restoration projects to MHHW elevation.

Sediment needed if 24,000 acres (~10,000 ha) of planned restoration are successful: Bayland sediment demand would increase by approximately 50% to 548 Mt of sediment if the 24,000 acres (~10,000 ha) of planned tidal marsh restoration throughout San Francisco Bay are completed (Figure 2.5). In the near term (by 2050), under this restoration scenario an additional ~119 Mt of sediment are needed, with over 80% (~97 Mt) needed to raise low-lying polders to local MHHW and the remaining 20% (~22 Mt) needed for the newly restored marsh to keep pace with SLR. In the long term (by 2100), an additional ~65 Mt of sediment are needed by the newly restored tidal marsh to continue to keep pace with SLR through the end of the century.

It is important to note that the habitat scenarios chosen in this report do not include plans to restore tidal flats, so the total sediment needed for tidal flats to keep pace with SLR does not change between habitat scenarios.

Figure 2.5. The total sediment needed for baylands to keep pace with 2.1 m (~6.9 ft) of SLR between 2010 and 2100 is approximately 363 Mt. If all planned tidal marsh restoration projects are successful and low-lying restoration sites are filled, an additional 184 Mt (increasing total sediment needed to 548 Mt) would be needed by the end of the century.

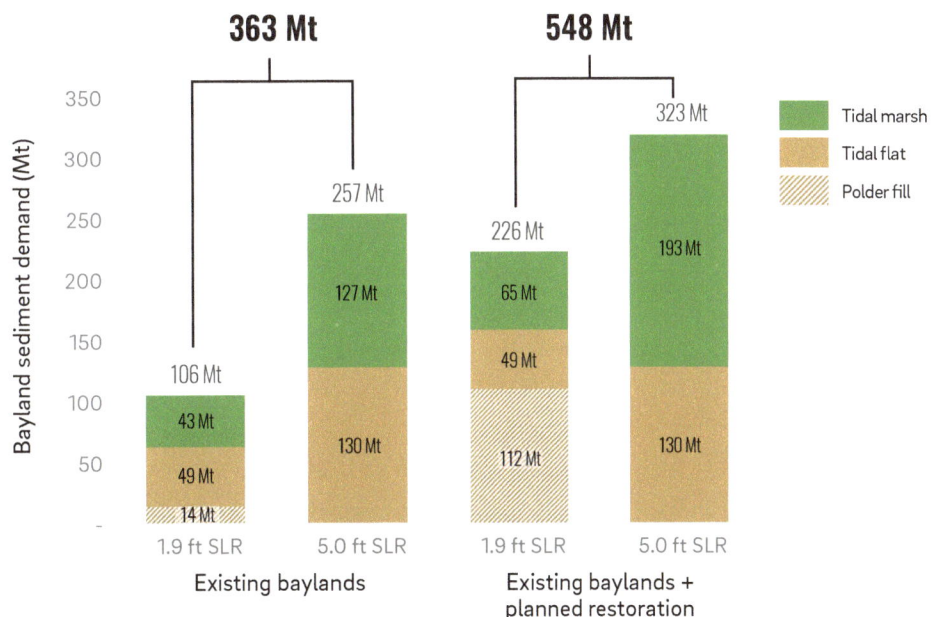

Future Baylands Sediment Demand at the Subembayment Scale

The baylands in San Pablo Bay account for over 40% of baywide sediment demand under both the existing baylands scenario and the existing baylands + planned restoration scenario (~152 Mt and 217 Mt respectively) (Figure 2.6). San Pablo Bay has the largest proportion of ongoing and planned tidal marsh restoration sites of all the subembayments. Cullinan Ranch, Sears Point/Dickson Ranch, Skaggs Island, Hamilton Wetlands, and various restoration projects at the mouth of the Napa River constitute more than 12,500 acres (~5,100 ha) of existing and planned conversion of subsided diked baylands to tidal marsh.

The baylands in South Bay have the next largest demand for sediment, accounting for approximately 25% of baywide sediment needed under the existing habitat scenario (~90 Mt) and approximately 22% under the existing baylands + planned restoration scenario (~120 Mt) (Figure 2.6). The proportion of ongoing and planned tidal marsh restoration sites in South Bay, which includes Inner Bair Island, Eden Landing Ecological Reserve Complex, and marshes at the Ravenswood Preserve, north of Dumbarton Bridge, comprise about 5,100 acres (~2,100 ha), or 17% of the baywide total.

The baylands located in Lower South Bay constitute about 16% of sediment needed baywide under the existing habitat scenario (59 Mt) and 22% under the restoration scenario (123 Mt) (Figure 2.6). This is slightly less than South Bay sediment demand. Lower South Bay has 9% more ongoing and planned marsh restoration area than South Bay, however, and about 26% (7,600 acres (~3,100 ha)) of the existing or planned restoration baywide. Additionally, Lower South Bay needs about 20% more sediment than South Bay to convert subsided restoration sites to MHHW. However, the differences in overall sediment demand between these subembayments are not due to tidal marshes, but rather to tidal flats. Based on the subembayment boundaries, South Bay has a more extensive shoreline that results in 1.5 times more tidal flat area than Lower South Bay. Because tidal flats are assumed to have a bulk density on average over two times greater than that of tidal marshes, the differences in tidal flat extent are driving the differences in overall baylands sediment demand between these two subembayments.

The baylands in Suisun Bay and Central Bay have similar demands for sediment under the existing baylands scenario, around 9% (33 Mt) and 8% (30 Mt) of baywide projections respectively (Figure 2.6). They have different drivers behind their similar sediment demands, however. Sustaining tidal marshes accounts for the majority of demands in Suisun Bay, while sustaining tidal flats accounts for the majority of demands in Central Bay. Under the existing baylands + planned restoration scenario, Suisun Bay accounts for about 10% (57 Mt) and Central Bay accounts for 5% (30 Mt) of baywide sediment demand. For these subembayments, the difference in sediment demand between the existing baylands and existing baylands + planned restoration

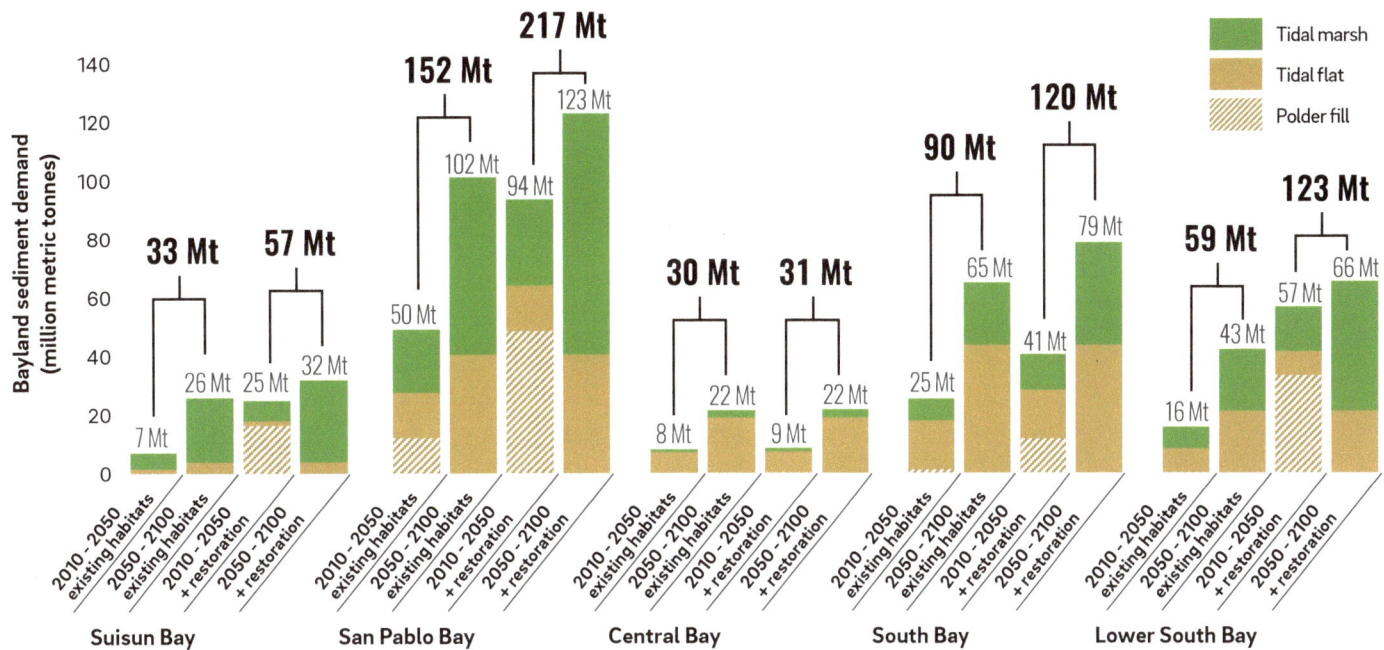

Figure 2.6. Near- and long-term bayland sediment demand by subembayment for existing baylands and existing baylands + planned restoration. San Pablo Bay has the highest bayland sediment demand compared to the rest of the Bay, with approximately 152 Mt of sediment needed for existing habitats to keep pace as sea level rises by 2.1 m (~6.9 ft) by the end of the century and an additional 65 Mt (totaling 217 Mt) if all planned tidal marsh restoration in San Pablo Bay is successful. Central Bay has the smallest bayland sediment demand of around 30 Mt of sediment with projections unchanged when analyzed for planned restoration since little space exists to restore tidal marsh within the baylands of Central Bay.

scenarios is due to their differences in planned marsh restoration. For Central Bay, where restoration opportunities are very limited, there is essentially no difference between scenarios. Relatively little restoration is planned for Central Bay. By contrast, about 4,000 acres (~1,600 ha) of restoration are planned for Suisun Bay, representing about 13% of ongoing and planned restoration baywide. In general, the restoration of brackish tidal marshes in Suisun Bay demands less inorganic sediment than restoration of saline marshes elsewhere in the study area. This is because autochthonous organic material contributes more to brackish marsh accretion than saline marsh accretion. Therefore, tidal marsh bulk density values used to calculate inorganic sediment demand are lesser for Suisun Bay than for the other subembayments (Table 2.1). For estimates of baylands sediment demand at the OLU scale, see Table 2.2 and Figure 2.7.

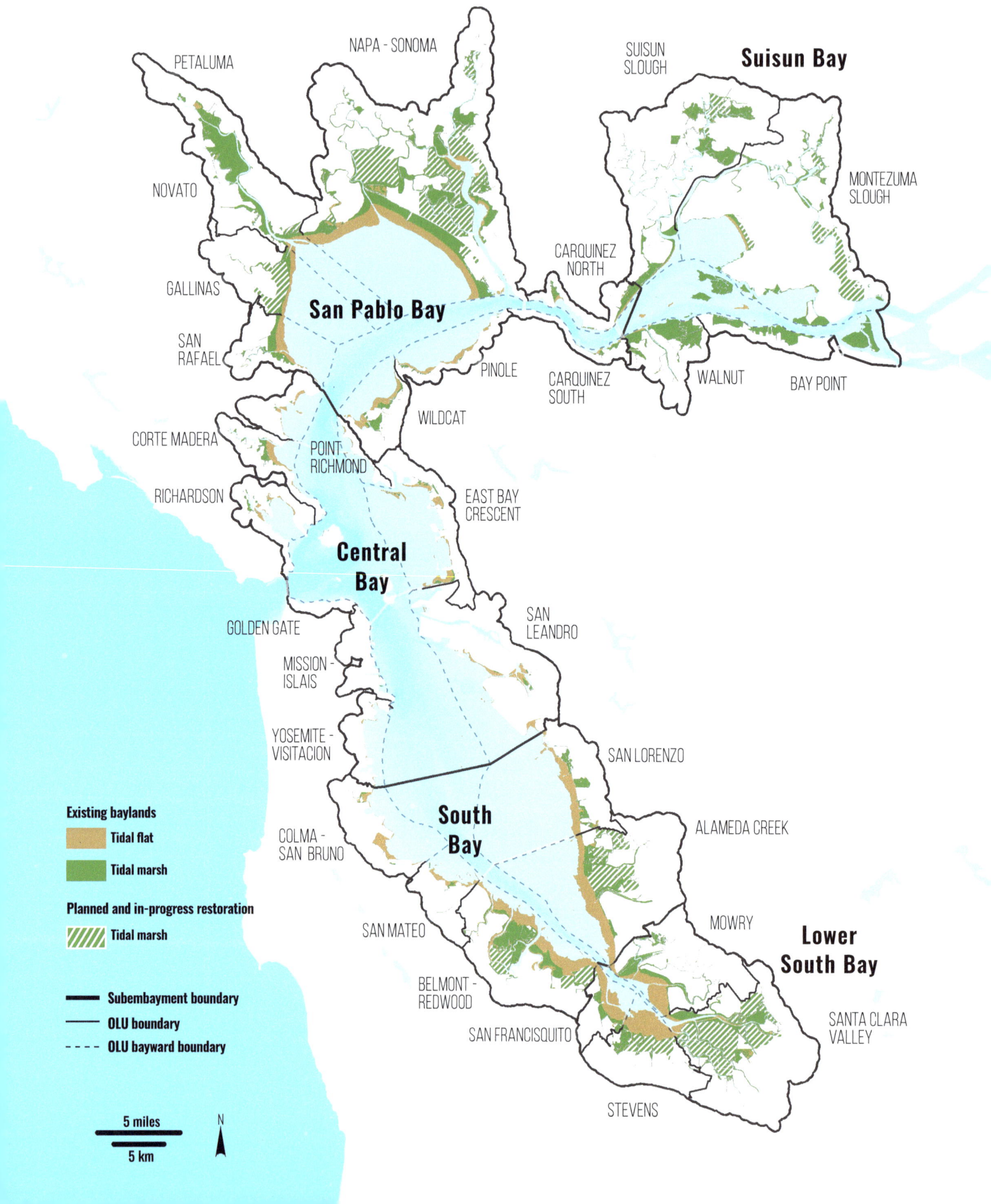

Suisun Bay

SUISUN
SLOUGH

MONTEZUMA
SLOUGH

CARQUINEZ
NORTH

CARQUINEZ
SOUTH

WALNUT

BAY POINT

PETALUMA

NAPA - SONOMA

NOVATO

GALLINAS

San Pablo Bay

SAN
RAFAEL

PINOLE

WILDCAT

CORTE MADERA

POINT
RICHMOND

RICHARDSON

EAST BAY
CRESCENT

**Central
Bay**

GOLDEN GATE

SAN
LEANDRO

MISSION -
ISLAIS

YOSEMITE -
VISITACION

SAN LORENZO

**South
Bay**

ALAMEDA CREEK

COLMA -
SAN BRUNO

SAN MATEO

MOWRY

**Lower
South Bay**

BELMONT -
REDWOOD

SANTA CLARA
VALLEY

SAN FRANCISQUITO

STEVENS

Existing baylands

Tidal flat

Tidal marsh

Planned and in-progress restoration

Tidal marsh

━━━ Subembayment boundary

─── OLU boundary

- - - OLU bayward boundary

5 miles

5 km

N

Table 2.2. Future baylands sediment demand of existing and planned baylands habitats for near- and long-term projections of SLR by OLU.

	OLU	Existing baylands		Existing baylands plus planned restoration	
		Near-term (1.9ft SLR) [Mt]	Long-term (5.0ft SLR) [Mt]	Near-term (1.9ft SLR) [Mt]	Long-term (5.0ft SLR) [Mt]
Suisun Bay	Islands*	1.2	4.4	1.4	4.5
Suisun Bay	Suisun Slough	1.9	7.5	3.7	8.6
Suisun Bay	Montezuma Slough	2.3	6.9	18.1	11.8
Suisun Bay	Bay Point	0.6	2.5	0.6	2.5
Suisun Bay	Walnut	1.2	4.8	1.2	4.8
San Pablo Bay	Gallinas	2.3	6.2	2.3	6.3
San Pablo Bay	Novato	1.5	4.1	13.8	9.0
San Pablo Bay	Petaluma	6.3	16.3	6.4	16.3
San Pablo Bay	Napa - Sonoma	34.8	63.2	66.6	79.8
San Pablo Bay	Carquinez North	1.1	2.9	1.1	2.9
San Pablo Bay	Carquinez South	0.3	0.8	0.5	0.8
San Pablo Bay	Pinole	1.2	3.1	1.2	3.1
San Pablo Bay	Wildcat	2.1	5.1	2.2	5.3
Central Bay	Islands**	<0.1	0.2	<0.1	0.2
Central Bay	Richardson	0.8	2.2	0.9	2.2
Central Bay	Corte Madera	1.6	4.0	1.8	4.1
Central Bay	San Rafael	0.7	1.8	0.7	1.8
Central Bay	Point Richmond	0.2	0.5	0.2	0.5
Central Bay	East Bay Crescent	2.5	6.6	2.5	6.6
Central Bay	San Leandro	2.2	5.8	2.2	5.8
Central Bay	Yosemite - Visitacion	0.3	0.7	0.3	0.7
Central Bay	Mission - Islais	<0.1	<0.1	<0.1	<0.1
Central Bay	Golden Gate	-	<0,1	-	<0.1
South Bay	San Lorenzo	4.1	10.9	4.3	11.2
South Bay	Alameda	9.1	21.1	16.6	28.8
South Bay	Belmont - Redwood	10.0	27.2	17.5	32.9
South Bay	San Mateo	0.9	2.3	0.9	2.3
South Bay	Colma - San Bruno	1.5	3.9	1.5	3.9
Lower South Bay	Mowry	4.7	12.9	4.7	12.9
Lower South Bay	Santa Clara Valley	4.9	12.8	36.7	31.1
Lower South Bay	Stevens	3.7	9.9	13.0	14.7
Lower South Bay	San Francisquito	2.6	7.0	2.8	7.2
		106.5	257.4	225.7	322.6

*Browns, Winters, Ryer, and Roe Islands are located in the deep channels of Suisun Bay and do not fall within a specified OLU

**Treasure Island and Angel Island are located in the deep channels of Central Bay and do not fall within a specified OLU

Figure 2.7. (left) Corresponding habitat footprints of the future mineral sediment demands summarized by Operational Landscape Units (OLUs) in Table 2.2. Dark hatched areas indicate breached tidal marsh restoration projects in the process of accreting to tidal marsh elevations as of 2009 (included in the existing baylands scenario). White hatched areas indicate areas acquired and slated for tidal marsh restoration as of 2015 (included in the existing baylands + planned restoration scenario). [Data Source: SFEI-ASC 2017b; Goals Project 2015]

Sediment Needed to Raise Low-lying Restoration Sites

Approximately 112 Mt of sediment are needed to raise all of the in-progress and planned tidal marsh restoration sites considered in this report to present-day, local MHHW elevations (Figures 2.8 and 2.9). To arrive at this estimate, we used tidal flat bulk densities averaged by subembayment, as detailed in Table 2.1, to convert soil volumes to mass of mineral sediment. For more information on methods used, see Appendix A.

Restoration sites located in San Pablo Bay account for 44% of the regional sediment demand for polder fill, about three-quarters of which is required for planned restoration. Lower South Bay, which contains many of the ponds slated for restoration under the South Bay Salt Pond Restoration Project, has the next largest need, equaling about 30% of regional sediment demands for polder fill. Nearly all of the polder fill needed in Lower South Bay (about 98%) is required for planned restoration.

Suisun Bay and South Bay make up a more moderate portion of regional sediment demands for polder fill—about 15% and 11% respectively. Central Bay has the lowest sediment demand for fill, with an estimated need of approximately 350,000 metric tonnes (t). The low demand in Central Bay reflects the limited opportunities for tidal marsh restoration in this subembayment, due to land development and steep Bay shore topography.

The estimate of 112 Mt of sediment needed to fill restoration sites reflects topobathymetry largely based on 2010 conditions. Thus, any progress made to raise low-lying restoration sites since 2010 is not accounted for in this estimate. Significant progress has been made since 2010 to fill some subsided sites for restoration. In particular, between 2010 and 2017, Hamilton and Cullinan Ranch projects received approximately 1.3 Mcy (~1.0 Mm^3) and 0.8 Mcy (~0.6 Mm^3) respectively, representing about 3% of the sediment demand for San Pablo Bay, and Montezuma Wetlands received about 4.5 Mcy (~3.4 Mm^3), representing about 18% of the demand for Suisun Bay (LTMS 2011, 2012, 2013, 2014, 2015, 2016, 2017, 2018). In addition, roughly ~1 Mcy (~0.8 Mm^3) of upland soils was placed at Inner Bair Island (BCDC 2017) before being breached in 2015 (USFWS 2015), representing roughly 5% of sediment demand for South Bay. §

Figure 2.8. Sediment demand to raise active and planned restoration sites to present-day, local MHHW elevations. Light brown indicates mass of mineral sediment needed to raise active restoration sites in existing baylands and dark brown indicates mass of mineral sediment needed to raise planned restoration sites in the existing baylands + planned restoration scenario. San Pablo Bay and Lower South Bay need the most polder fill compared to the other subembayments, with the majority needed for planned tidal marsh restoration.

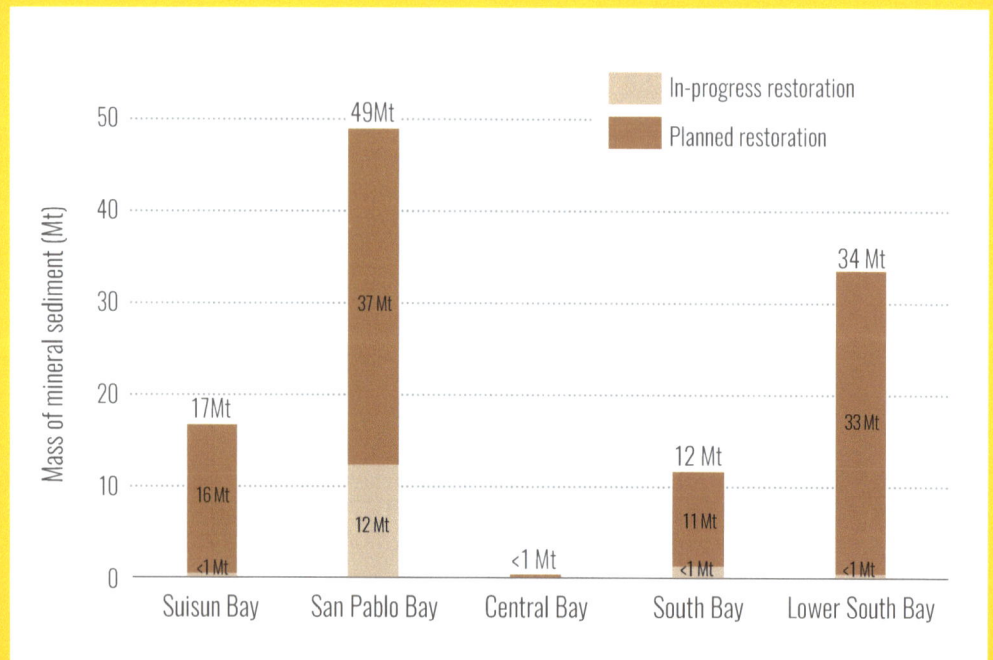

Suisun Bay

San Pablo Bay

NAPA - SONOMA

PETALUMA

SKAGGS ISLAND

NAPA PONDS

CULLINAN RANCH WETLANDS

NOVATO

SUISUN SLOUGH

MONTEZUMA SLOUGH

MONTEZUMA WETLANDS

BEL MARIN KEYS V

HAMILTON WETLAND

GALLINAS

SAN RAFAEL

CARQUINEZ NORTH

CARQUINEZ SOUTH

WALNUT

BAY POINT

PINOLE

CORTE MADERA

POINT RICHMOND

WILDCAT

RICHARDSON

EAST BAY CRESCENT

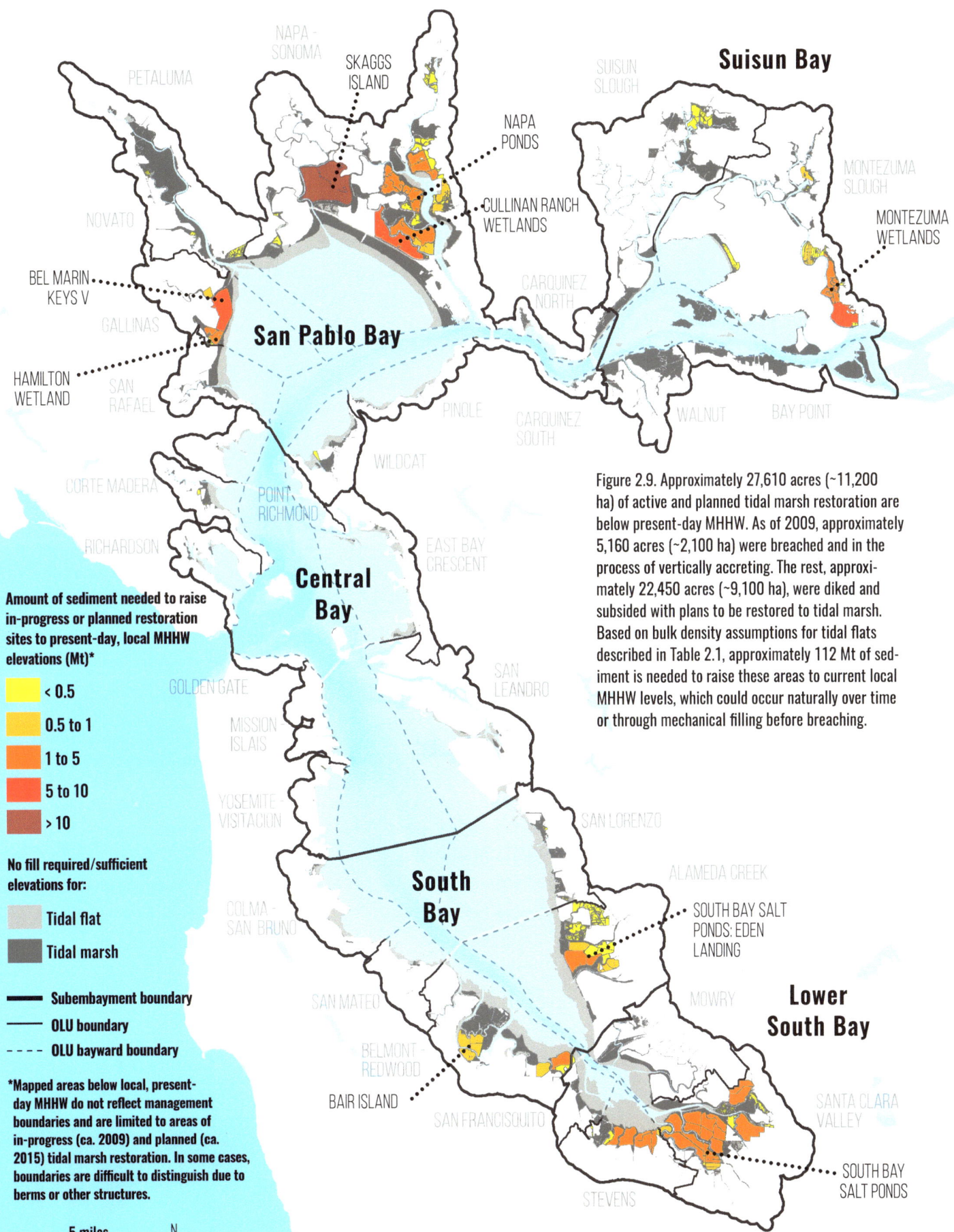

Figure 2.9. Approximately 27,610 acres (~11,200 ha) of active and planned tidal marsh restoration are below present-day MHHW. As of 2009, approximately 5,160 acres (~2,100 ha) were breached and in the process of vertically accreting. The rest, approximately 22,450 acres (~9,100 ha), were diked and subsided with plans to be restored to tidal marsh. Based on bulk density assumptions for tidal flats described in Table 2.1, approximately 112 Mt of sediment is needed to raise these areas to current local MHHW levels, which could occur naturally over time or through mechanical filling before breaching.

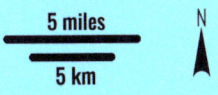

Central Bay

SAN LEANDRO

Amount of sediment needed to raise in-progress or planned restoration sites to present-day, local MHHW elevations (Mt)*

- 🟨 < 0.5
- 🟧 0.5 to 1
- 🟧 1 to 5
- 🟥 5 to 10
- 🟫 > 10

GOLDEN GATE

MISSION ISLAIS

YOSEMITE VISITACION

SAN LORENZO

ALAMEDA CREEK

No fill required/sufficient elevations for:

- ⬜ Tidal flat
- ⬛ Tidal marsh

— **Submembayment boundary**

— **OLU boundary**

--- **OLU bayward boundary**

South Bay

COLMA - SAN BRUNO

SOUTH BAY SALT PONDS: EDEN LANDING

MOWRY

Lower South Bay

SAN MATEO

***Mapped areas below local, present-day MHHW do not reflect management boundaries and are limited to areas of in-progress (ca. 2009) and planned (ca. 2015) tidal marsh restoration. In some cases, boundaries are difficult to distinguish due to berms or other structures.**

BELMONT REDWOOD

BAIR ISLAND

SAN FRANCISQUITO

SANTA CLARA VALLEY

SOUTH BAY SALT PONDS

STEVENS

5 miles

5 km

N

Sensitivity Analysis: How changes in bulk density estimates affect overall demand results

The bulk densities used to convert soil volumes to sediment mass linearly affect the estimates of sediment demand. For example, a 10% increase in bulk density of tidal marsh sediment yields a 10% increase in tidal marsh sediment demand. In order to gauge the variability of the mineral component of dry bulk density within soil cores analyzed and thus the variability of overall sediment demand, we used unpublished data for 0.02-m (~0.79-in) sections of cores provided by John Callaway (pers. comm.). To provide a measure of this variability, we calculated the degree to which sediment demands for tidal marsh would change using +/- one standard deviation of the mean bulk densities in the top 0.2 m (~0.7 ft) and ~0.4 m (~1.4 ft) of cores for near- and long-term SLR demand estimates respectively. This assessment was limited to San Pablo Bay because of the large number of cores and the range of marsh types within the salinity gradient sampled relative to the other subembayments (i.e., 23 soil cores; China Camp (n = 6), Coon Island (n = 11), Petaluma (n = 6)). Results indicate that the overall change in sediment demand for San Pablo Bay tidal marshes is +24% and -21% for both baylands scenarios when using +/- one standard deviation of measured bulk densities in soil cores respectively (Table 2.3). This equates to a change of about +20 Mt to -17 Mt for the existing baylands scenario and +26 Mt to -23 Mt for the existing baylands + planned restoration scenario. §

Table 2.3. Variability of the mineral component of dry bulk density from unpublished data provided by John Callaway (pers. comm.) for the top 0.2 m (~0.7 ft) and ~0.4 m (~1.4 ft) of soil cores for San Pablo Bay. Overall sediment demand by tidal marsh in San Pablo Bay would change by +24% and -21% when using +/- one standard deviation of the average mineral component of dry bulk densities to inform sediment demand calculations.

Assumed mineral component of dry bulk density	Mineral component of dry bulk density within the top 20 cm of cores (t/m^3) for San Pablo Bay tidal marshes [used to calculate near-term sediment demand estimates]	Mineral component of dry bulk density within the top 44 cm of cores (t/m^3) for San Pablo Bay tidal marshes [used to calculate long-term sediment demand estimates]	Total sediment needed for existing baylands to keep pace with 6.9 ft of SLR through 2100 (Mt)	Total sediment needed for existing baylands + planned restoration to keep pace with 6.9 ft of SLR through 2100 (Mt)
Average	0.40	0.43	82	112
+1 standard deviation	0.50	0.52	102	89
-1 standard deviation	0.30	0.34	65	138

Bay Sediment Supply Analysis

Methods and Assumptions

Future Delta Sediment Supply

Estimates of future sediment supply to the Bay from the Delta were derived from modeled future Sacramento River sediment loads. As part of the USGS CASCaDE project, Stern et al. (2020) modeled annual total sediment load (suspended load and bedload) for Sacramento River at Freeport for WY2017–2100 for a suite of future climate scenarios including CESM1-BGC RCP 8.5 (the wetter future in this study) and HadGEM2-CC RCP 8.5 (the drier future in this study) (Figure 2.10). We converted these sediment loads to future annual sediment load to the Bay at Mallard Island using a regression equation relating modeled historical annual total sediment load for Sacramento River at Freeport (from Stern et al. 2016) and calculated historical Mallard Island annual suspended sediment load (from Schoellhamer et al. 2018) for WY1995–2008 (Figure 2.11). Based on the discussion in Schoellhamer et al. (2018) regarding the overall reduction in bedload transport to Mallard Island and from sand mining shipping channel dredging, the future average annual bedload transport to the Bay at Mallard Island was assumed to be zero.

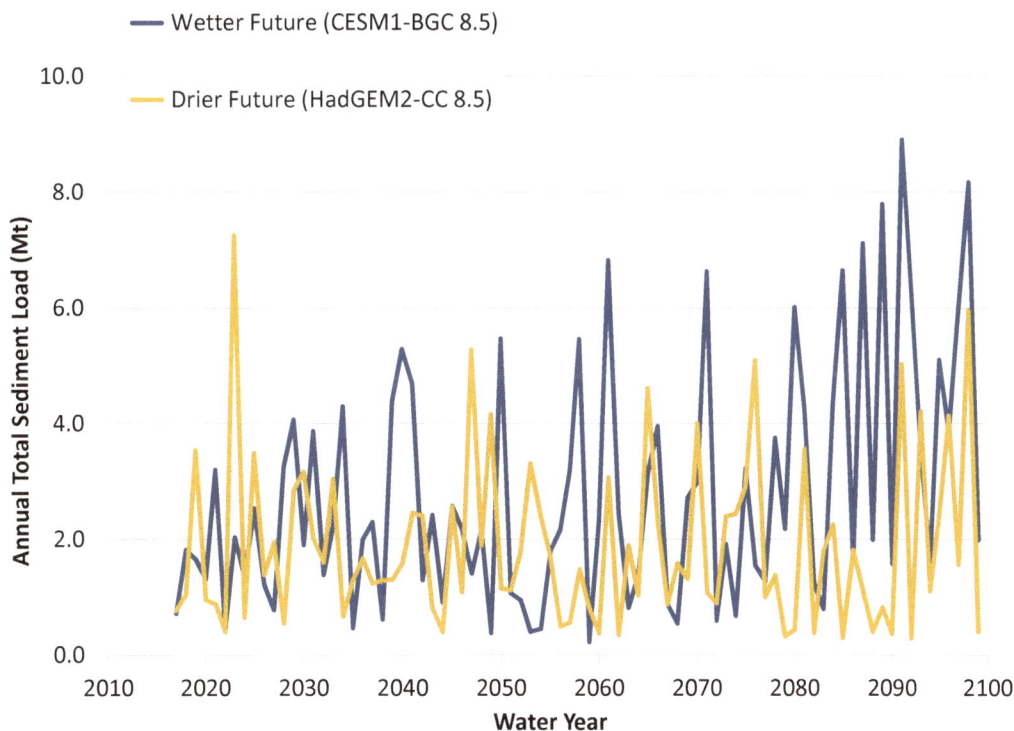

Figure 2.10. Modeled total sediment load for Sacramento River at Freeport for the wetter future (CESM1-BGC 8.5) and drier future (HadGEM2-CC 8.5). [data source: Stern et al. 2020]

Figure 2.11. Correlation between calculated suspended sediment load at Mallard Island and modeled total Sediment load for Sacramento River at Freeport (WY1995–2008). [data source: Stern et al. 2016 and Schoellhamer et al. 2018].

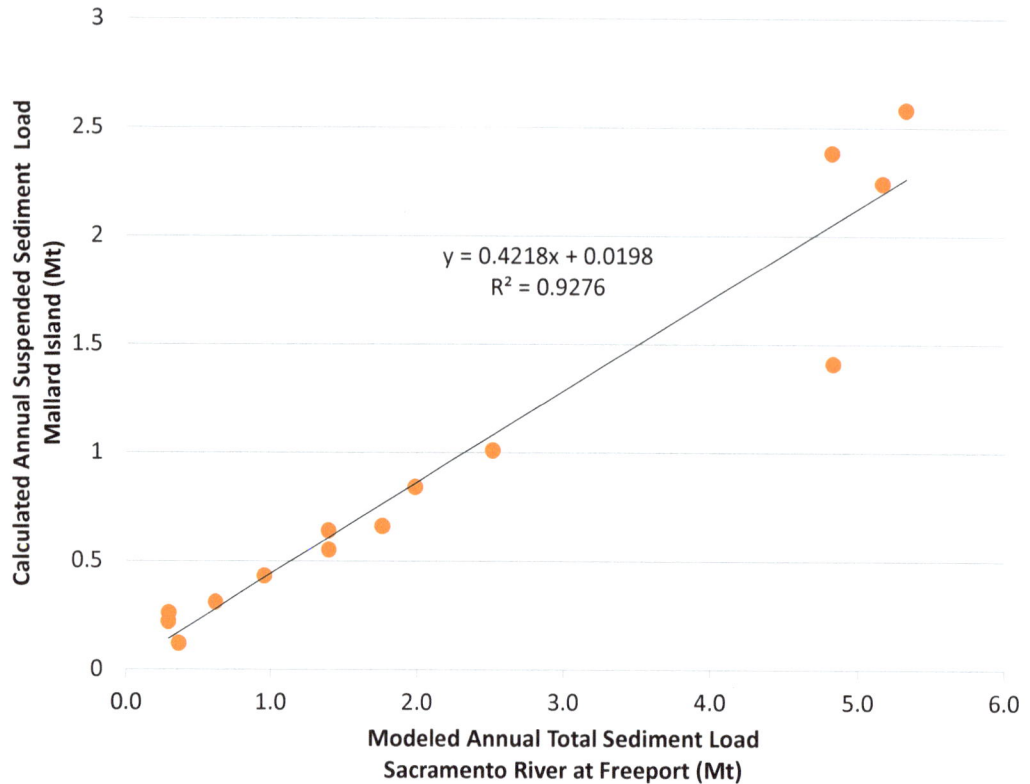

$$y = 0.4218x + 0.0198$$
$$R^2 = 0.9276$$

Future Tributary Sediment Supply

Focus Tributaries

An intensive analysis of future sediment supply to the study area for both climate scenarios was conducted for several focus Bay tributaries, and these results were then applied to all other tributaries to arrive at a comprehensive assessment of future tributary sediment supply (Figure 2.12). Nine focus tributaries were selected for intensive analysis: Alameda Creek, Corte Madera Creek, Guadalupe River, Napa River, San Francisquito Creek, San Leandro Creek, San Lorenzo Creek, Sonoma Creek, and Walnut Creek. These tributaries were selected because they represent a range of watershed area, location in the study area, and almost half of the historical annual sediment loads for all Bay tributaries (per SFEI-ASC 2017a and Schoellhamer et al. 2018). In addition, Alameda, Napa, Sonoma, and Walnut were all included because they are estimated to be by far the highest sediment producing tributary watersheds in the study area, and therefore will play a large role in future sediment supply to the Bay.

For each focus tributary, future annual sediment load to the Bay was determined for the wetter and drier futures at approximately head of tide, or the inland extent of mean higher high water, using a combination of historical sediment rating curves and modeled future runoff.

Historical sediment rating curves: Historical sediment rating curves were developed by combining estimates of annual total sediment load and annual runoff. Annual total sediment loads for the recent past (WY1960–2016) came from existing analyses that

Napa

Sonoma

Corte
Madera

Walnut

San Leandro

San Lorenzo

Alameda

10 miles
10 km
N

San
Francisquito

Guadalupe

Figure 2.12. Map showing the
Focus Tributaries used in this
study.

used field measurements of sediment flux to estimate sediment load for gaged and ungaged Bay tributaries (see SFEI-ASC 2017a and Schoellhamer et al. 2018). Annual runoff volumes for this period for each focus tributary were provided by Lorraine Flint from the USGS, who ran the Basin Characterization Model (BCM) with historical precipitation to calculate monthly runoff, which was then compiled into annual runoff (see Flint et al. 2013 and Flint et al. 2020 for modeling methods and assumptions). A regression equation relating annual runoff and annual total sediment load (or a historical sediment rating curve) for each focus tributary was then determined. We focused on the period WY1960–2016 to ensure that the rating curves captured a wide range of wet and dry water years.

Modeled future runoff: Future runoff volumes for each focus tributary for WY2017–2100 were provided by Lorraine Flint, who ran the BCM with downscaled monthly precipitation for the wetter and drier climate scenarios to calculate future monthly runoff, which was then compiled into future annual runoff (see Flint and Flint 2012 and Flint et al. 2020 for modeling methods and assumptions).

Future annual sediment load: Future annual sediment loads for each focus tributary for the wetter and drier climate scenarios were then determined by combining future annual runoff volumes with the historical sediment rating curve.

Non-focus Tributaries

Future annual sediment loads for the remaining 338 tributary watersheds that drain directly to the Bay were determined based on the findings from the nine focus tributary watersheds. This began by assessing the magnitude of change of average annual sediment load for the focus watersheds between the recent past (WY1995–2016) and the wetter and drier future scenarios. We then averaged the loads for the focus watersheds north of the Golden Gate (Walnut, Corte Madera, Sonoma, Napa) and south of the Golden Gate (San Lorenzo, San Leandro, Alameda, San Francisquito, Guadalupe) to determine average annual sediment load change "multipliers" for both time periods for the wetter and drier climate scenarios for use with non-focus tributaries north and south of the Golden Gate, respectively. The multipliers were then combined with non-focus tributary watershed average annual sediment loads for the recent past (WY1995–2016) to arrive at future (WY2017–2100) average annual sediment loads for the wetter and drier climate scenarios.

Future Sediment Flux Between Subembayments and at the Golden Gate

Between Subembayments

Assessing the annual net flux between Bay subembayments is essential for understanding the portion of the annual local sediment supply that remains within individual subembayments and could therefore be available for local bayland deposition, and the portion that is transported out of subembayments and could therefore be available for bayland deposition elsewhere in the study area. For the recent past, annual net suspended sediment flux at Benicia bridge (the boundary between Suisun Bay and San Pablo Bay) has ranged from 21 Mt to San Pablo Bay in WY1998 (Ganju and Schoellhamer 2006)

Table 2.4. Annual Delta outflow and sediment flux at Bay subembayment boundaries (WY1997–1998, 2002–2016)

WY	Delta Outflow at Mallard Island (Mm³)[a]	Benicia Bridge Sediment Flux (Mt)[a,b]	Dumbarton Bridge Sediment Flux (Mt)[c]	Near Richmond Bridge Sediment Flux (Mt)[d]	Near Bay Bridge Sediment Flux (Mt)[e]
1997	42,300	5.1			
1998	53,600	20.9		2.5	1.0
2002	11,343	-2.1		2.1	1.8
2003	17,356	1.3			
2004	18,610	0.9			
2005	19,046	1.0			
2006	51,254	10.4		3.6	0.8
2007	7,643	-2.9			
2008	8,312	-2.7			
2009	8,443	-2.5	0.3		
2010	12,539	-2.0	0.2		
2011	33,235	4.6	-0.15		
2012	9,940	-2.3		2.0	1.3
2013	11,299	-2.1	-0.02		
2014	5,290	-3.0	0.6		
2015	7,687	-2.0	0.6		
2016	14,048	-0.9	0.3		

[a] source: Ganju and Schoellhamer 2006 (1997-1998), Schoellhamer et al. 2018 (2002-2016)
[b] negative values indicate a landward flux into Suisun Bay, positive values indicate a seaward flux into San Pablo Bay
[c] source: Livsey et al. 2020; negative values indicate a seaward flux into South Bay, positive values indicate a landward flux into Lower South Bay
[d] source: Delta Modeling Associates 2015; negative values indicate a landward flux to San Pablo Bay, positive values indicate a seaward flux to Central Bay
[e] source: Delta Modeling Associates 2015; negative values indicate a landward flux to South Bay, positive values indicate a seaward flux to Central Bay

to 3 Mt to Suisun Bay in WY2014 (Schoellhamer et al. 2018) (see Table 2.4). Annual net suspended sediment flux at Dumbarton Bridge (the boundary between South Bay and Lower South Bay) has ranged recently from 0.02 Mt to South Bay in WY2013 to 0.6 Mt to Lower South Bay in WY2014 and WY2015 (Livsey et al. 2020). These measurements show a net flux direction downstream (bayward) from Suisun Bay to San Pablo Bay, and upstream (landward) from South Bay to Lower South Bay (see Table 2.4). At both locations, the flux magnitude and direction varied as a function of water year type, as indicated by Delta outflow volume (i.e., high outflow volumes are associated with the wettest water years). Near the Richmond Bridge (the boundary between San Pablo Bay and Central Bay) and near the Bay Bridge just north of the Central Bay–South Bay boundary, modeling of the recent past suggests the net annual sediment flux magnitude varies with water year type but the net annual flux direction is constant (see Table 2.4). Delta Modeling Associates (2015) showed that annual net sediment flux near the Richmond Bridge was consistently towards Central Bay for WY1998, 2002, 2006, and 2012, with the annual flux varying between 1.4 and 2.4 Mcy (~1.1 and ~1.8 Mm³) (2.0 and 3.6 Mt/yr)[1] and the fluxes being highest for the wettest water years (WY1998 and 2006). Sediment flux modeling near the Bay Bridge also showed a consistent annual net sediment flux direction towards Central Bay, but the annual fluxes varied between 0.6 and 1.2 Mcy (~0.5 and ~0.9 Mm³) (0.8 and 1.8 Mt/yr)[1] and were the highest for the driest water years modeled (WY2002 and 2012) (Delta Modeling Associates 2015).

---

1   Annual mass calculated by multiplying annual volume (converted to m³) by a bulk density of 1,300 kg/m³ and a porosity of 0.7. See Delta Modeling Associates (2015) for more details.

The complex spatial and temporal nature of sediment transport dynamics among the subembayments prohibited using the sediment flux information from the recent past to predict future sediment flux between embayments. For Benicia Bridge and Dumbarton Bridge, we concluded that it was not possible to determine the magnitude nor the direction of future annual sediment flux without new numerical modeling that is beyond the scope of this study. For the Richmond Bridge and the Bay Bridge, we were not able to determine the magnitude of future annual flux but we assumed that the net direction of future annual flux would continue to be towards Central Bay.

At the Golden Gate

The sediment flux through the Golden Gate is critical for understanding the mass of sediment in the study area that is available for bayland deposition. Recent numerical modeling of Golden Gate suspended sediment flux for WY2004–2010 shows a net export out of the Bay, with an average annual net sediment flux of 1.2 Mt/yr (Erikson et al. 2013). Schoellhamer et al. (2005) report a similar average annual net sediment flux out the Golden Gate for WY1995–2002. We determined that recent flux data could not be used to estimate future flux amounts. However, we assumed that the direction of future average annual net sediment flux would continue to be out the Golden Gate.

Central Bay (Landsat imagery courtesy NASA, April 2013)

FUTURE SEDIMENT SUPPLY:
Key Assumptions and Considerations

- There are a variety of factors associated with climate, land use, water management changes, and large-scale Delta island restoration through intentional or unintentional breaching, as well as the potential for catastrophic failure of Delta levees, that could affect future sediment loading to the Bay.

- This analysis assumes that the historical relationships between annual runoff and annual sediment load at the focus tributaries remain the same for the 21st century. There are variety of factors that could alter future tributary flow-sediment relationships, including changes to precipitation patterns, land cover type and distribution, wildfire frequency and severity, and flow management.

- The Mission-Islais and Golden Gate OLUs were not included in the future tributary sediment supply analysis due to lack of historical rating curves.

- The ratio of future to historical average annual sediment load for the focus tributaries is assumed to be applicable to all other Bay tributaries. However, variability in runoff dynamics and sediment loads associated with local climatic and land use factors could cause big differences in ratios between focus tributaries and neighboring tributaries.

- The future net sediment flux direction between San Pablo Bay and Central Bay, South Bay and Central Bay, and Central Bay the Pacific Ocean is assumed to the be the same as historical conditions and does reflect changes that could result from sea-level rise and changing sediment delivery caused by climate and land use changes.

- This analysis assumes that the dominant sediment sources for bayland vertical accretion are from the Delta and Bay tributaries and does not account for sediment supplied from eroding deep channels, shallows, tidal flats, tidal channels, and shorelines. Although these features can collectively supply a considerable amount of sediment, the lack of data about their future erosion rates and transport of the eroded sediment to tidal and marshes under a rising sea level precludes accounting for their erosion in the estimates of bayland sediment supply and demand.

- The effects of increased variability in annual rainfall, especially the expected increase in the duration of droughts, is not fully captured in this analysis. Recent studies suggest that droughts and deluges have delayed effects on tidal marsh accretion rates, and that droughts have a greater effect than deluges. Decreases in sediment delivery to marshes and increases in marsh salinity are associated with droughts, which therefore can cause shifts in marsh vegetation composition, especially in brackish areas, which in turn can affect the contribution of vegetation to marsh accretion. §

Results and Discussion

Future Sediment Supply from the Delta

For the wetter future, the annual sediment load to the Bay from the Delta varies between 0.12 and 3.8 Mt and shows an upward trend from WY2017–2100 (Figure 2.13). The drier future has similar minimum and maximum annual sediment loads (0.14 and 3.1 Mt, respectively) but no upward or downward trend over time. For the wetter future, the average annual sediment load at mid-century (WY2028–2049) is approximately 45% higher than the recent past (WY1995–2016) and the average annual sediment load at the end of the century (WY2078-2099) is more than 150% higher than the recent past (Figure 2.14) (Figure 2.14). For the drier future, the average annual sediment load at mid-century is approximately 15% higher than the recent past and the average annual sediment load at the end of the century is less than 10% higher than the recent past. These findings suggest that both the wetter future and drier future scenarios could result in a modest increase in average annual sediment supply from the Delta to the study area by mid-century, but that the wetter future could result in a much higher sediment supply to the study area in the second half of the 21st century.

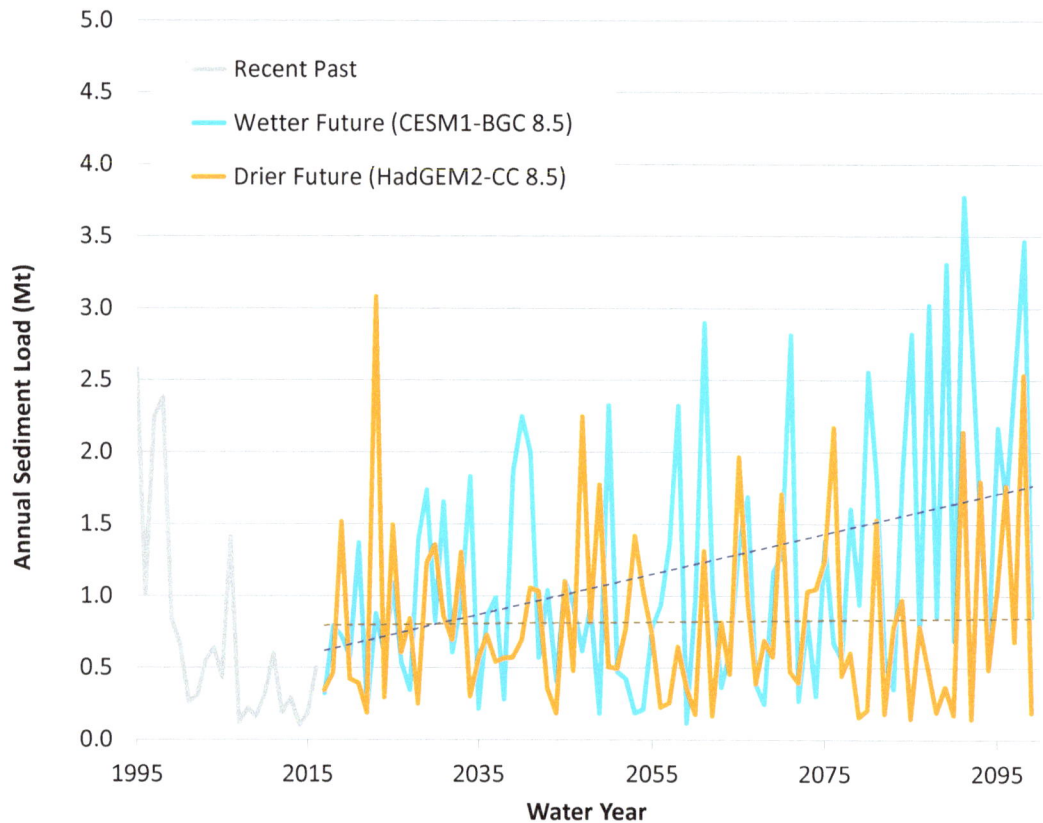

Figure 2.13. Annual sediment load coming into the Bay from the Delta at Mallard Island for the recent past (WY1995–2016) and the wetter future (CESM1-BGC 8.5) and drier future (HadGEM2-CC 8.5). [Data source: Schoellhamer et al. 2018, Stern et al. 2020]

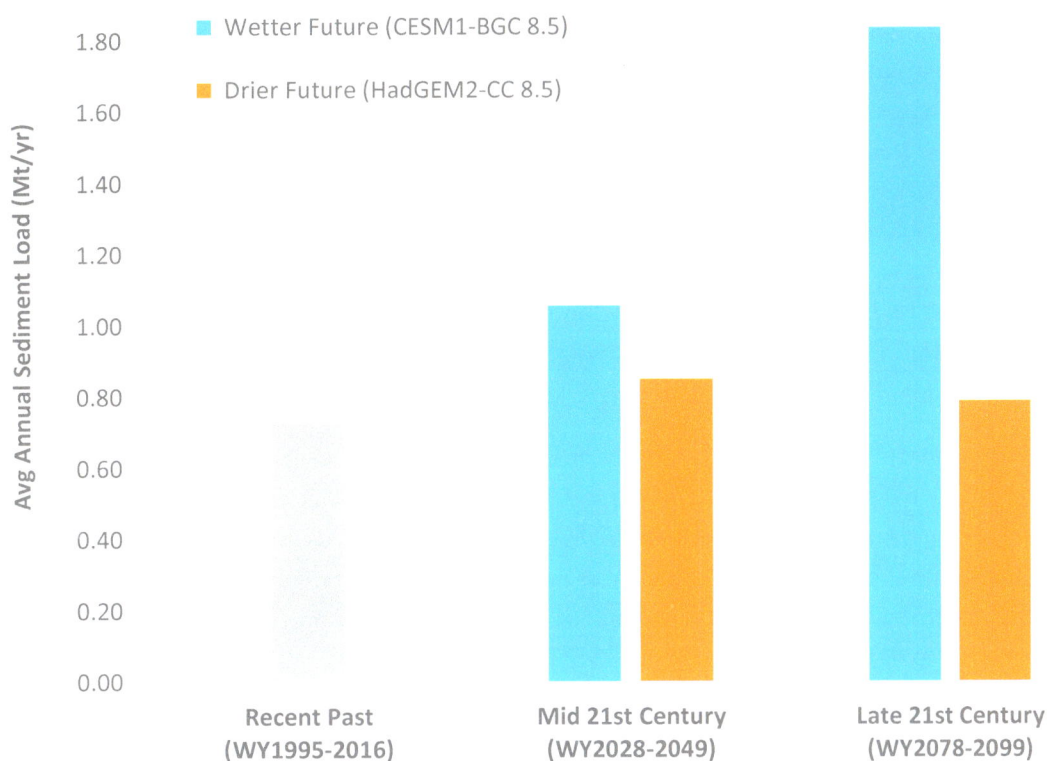

Figure 2.14. Average annual sediment load coming into the Bay from the Delta at Mallard Island for the recent past (WY1995–2016) and the wetter future (CESM1-BGC 8.5) and drier future (HadGEM2-CC 8.5) for mid-century (WY2028-2049) and late century (WY2078–2099). [Data source: Schoellhamer et al. 2018, Stern et al. 2020]

Future Sediment Supply from Focus Tributaries

For the wetter future, the degree of change in average annual sediment supply varies among the focus tributaries due to regional differences in precipitation and runoff changes. However, all focus tributaries show higher average annual sediment loads at the end of the century compared to the recent past, with the historically dominant sediment-supplying tributaries in North Bay showing the greatest increase (Figure 2.15). In Lower South Bay, Guadalupe River has a mid-century average annual sediment load that is approximately 20% higher than the recent past and an end of century average annual load that is almost 200% higher than the recent past. San Francisquito Creek, however, has a mid-century average annual sediment that is approximately 10% higher than the recent past and an end of century average annual load that is 100% higher than the recent past. In South Bay, Alameda Creek and San Lorenzo Creek have mid-century average annual sediment loads that are approximately 30% lower than the recent past but end of century average annual loads that are more than 20% higher than the recent past. In Central Bay, San Leandro Creek's mid-century average annual sediment load is approximately 10% less than the recent past but the end of century average annual load is approximately 60% higher than the recent past. Conversely, Corte Madera Creek in Central Bay has a mid-century average annual sediment load that is 100% higher than the recent past and an end of century average annual load that is approximately 300% higher than the recent past. In the northern portion of the Bay, Sonoma Creek has a mid-century average annual sediment load that is approximately 20% lower than the recent past, Walnut Creek has a mid-century average annual load that is approximately 10% higher than the recent past, and Napa River has a mid-century average annual load that is approximately 60% higher than the recent past. At the end of the century, the Sonoma Creek average annual sediment load is approximately 60% higher than the recent past, the Walnut Creek average annual load is approximately 120% higher than the recent past, and the Napa River average annual load is approximately 280% higher than the recent past. This equates to the Sonoma Creek average annual sediment load at the end of the century being 0.1 Mt/yr higher than the recent past, the Walnut Creek load being 0.2 Mt/yr higher, and the Napa River load being 0.5 Mt/yr higher.

For the drier future, most focus tributaries show average annual sediment loads that are either similar to or lower than the recent past, with Corte Madera Creek and Napa River being the exceptions (Figure 2.15). In Lower South Bay, Guadalupe River and San Francisquito Creek both have mid-century average annual sediment loads that more than 40% lower than the recent past and end of century average annual loads that are more than 20% lower than the recent past. In South Bay, Alameda Creek and San Lorenzo Creek both have mid-century and end of century average annual sediment loads that are 70% lower than the recent past. In Central Bay, San Leandro Creek's mid-century and end of century average annual sediment load is 40% lower than the recent past. Conversely, Corte Madera Creek's mid-century average annual sediment load is similar to the recent past and end of century average annual load that is 60% higher than the recent past. The increase in Corte Madera Creek average annual load at the end of the century is driven largely by one wet water year with a high annual sediment load (see Appendix B). In the northern portion of the Bay, Sonoma Creek and Walnut Creek both have mid-century average annual sediment loads that are approximately 50% lower than the recent past and end of century average annual loads that are 40% lower than the recent past. Napa River, however, has a mid-century average annual sediment load that is only 20% lower than the recent past and an end of century average annual load that is 40% higher than the recent past. This increase in Napa average annual load at the end of the century is driven by two wet water years with very high annual loads (see Appendix B).

Combined Future Sediment Supply Results

Baywide (Delta and Bay Tributaries)

The compilation of the future supply of sediment to the Bay from the Delta and all 347 Bay tributaries for the full WY2010–2100 time period assessed for bayland demand shows the wetter future would result in 75% more sediment to the Bay than the drier future (280 Mt compared to 160 Mt) (Figure 2.16). The Delta contribution to the total sediment supply is slightly higher in the drier future than the wetter future (44% compared to 36%). For comparison, the Delta is thought to have supplied approximately 37% of the total sediment supply to the Bay in the recent past (WY1995–2016) (Schoellhamer et al. 2018). When considering sediment supply at the multi-decadal scale, the wetter and drier futures show similarities in the near term and considerable differences later in the

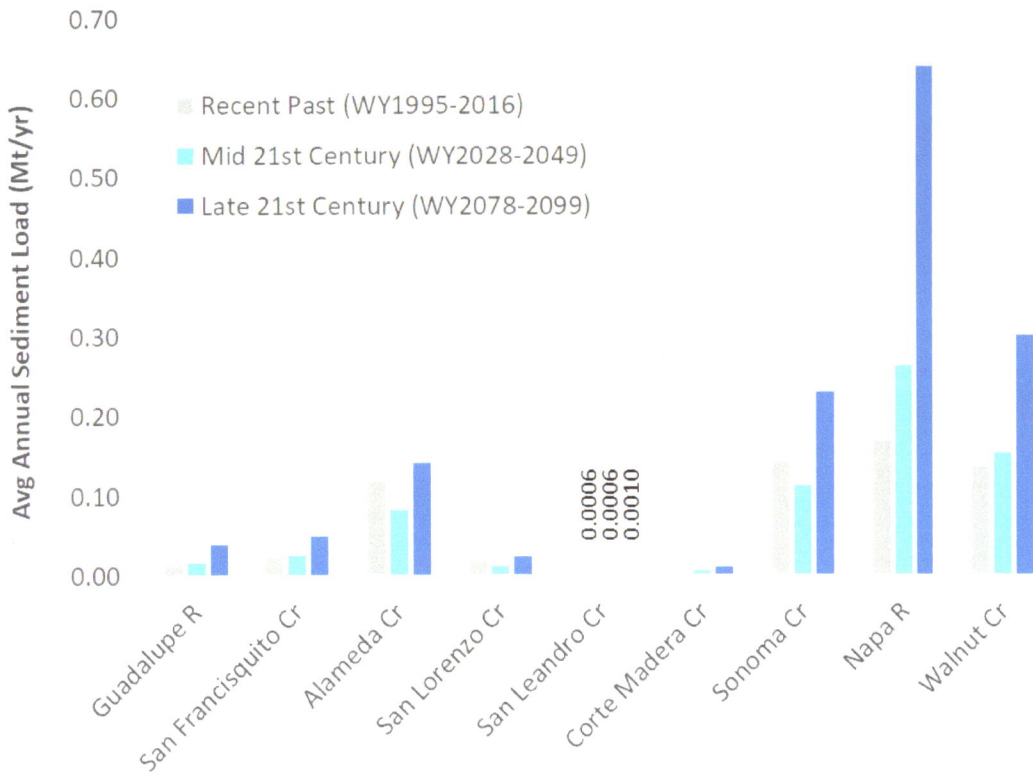

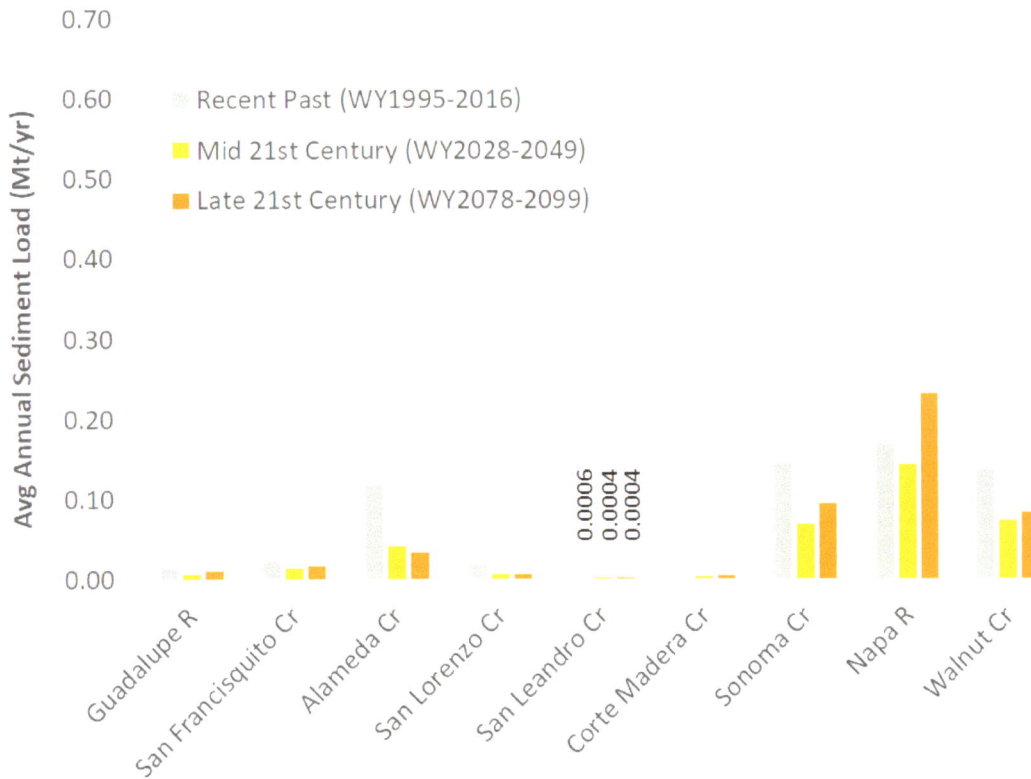

Figure 2.15. Modeled average annual sediment load for the focus tributaries for the recent past (WY1995–2016), mid-century (WY2028–2049) and late century (WY2078–2099) for the wetter future (CESM1-BGC 8.5, top) and drier future (HadGEM2-CC 8.5, bottom). See Appendix B for the focus tributary sediment rating curves and time series of modeled annual runoff and sediment load.

Figure 2.16. Total Sediment Supply to the Bay from the Delta (striped colors) and Bay Tributaries (solid colors) for the wetter future (CESM1-BGC 8.5) and drier future (HadGEM2-CC 8.5) for WY2010–2050 and WY2050–2100.

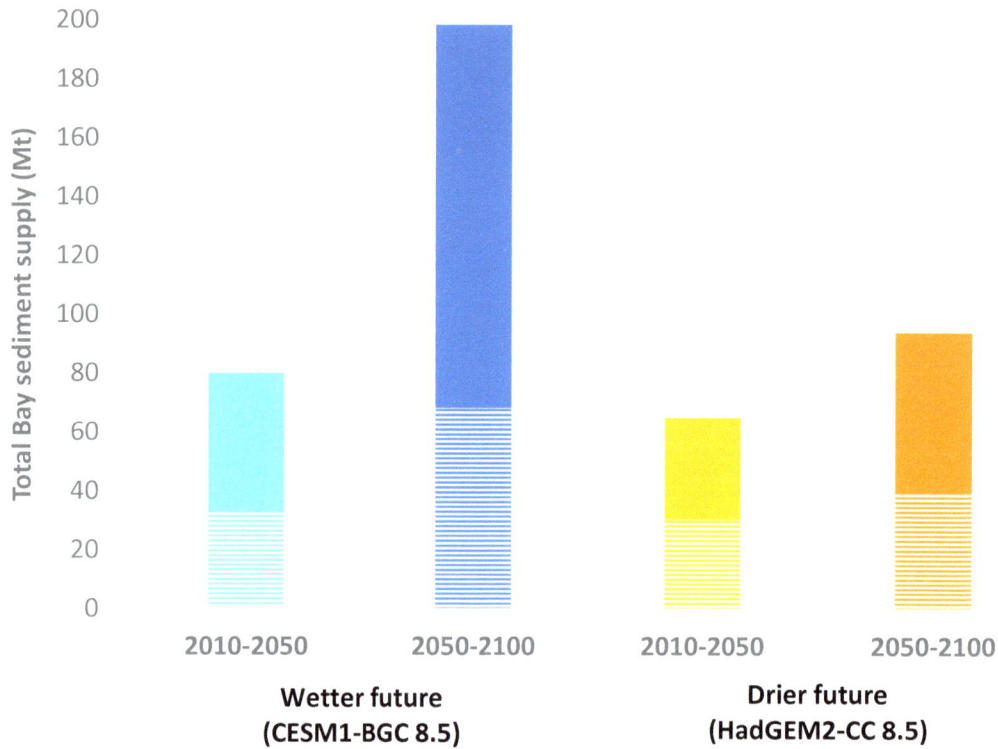

Figure 2.17. Total tributary sediment supply to each subembayment for the wetter future (CESM1-BGC 8.5) and drier future (HadGEM2-CC 8.5) for WY2010–2050 and WY2050–2100.

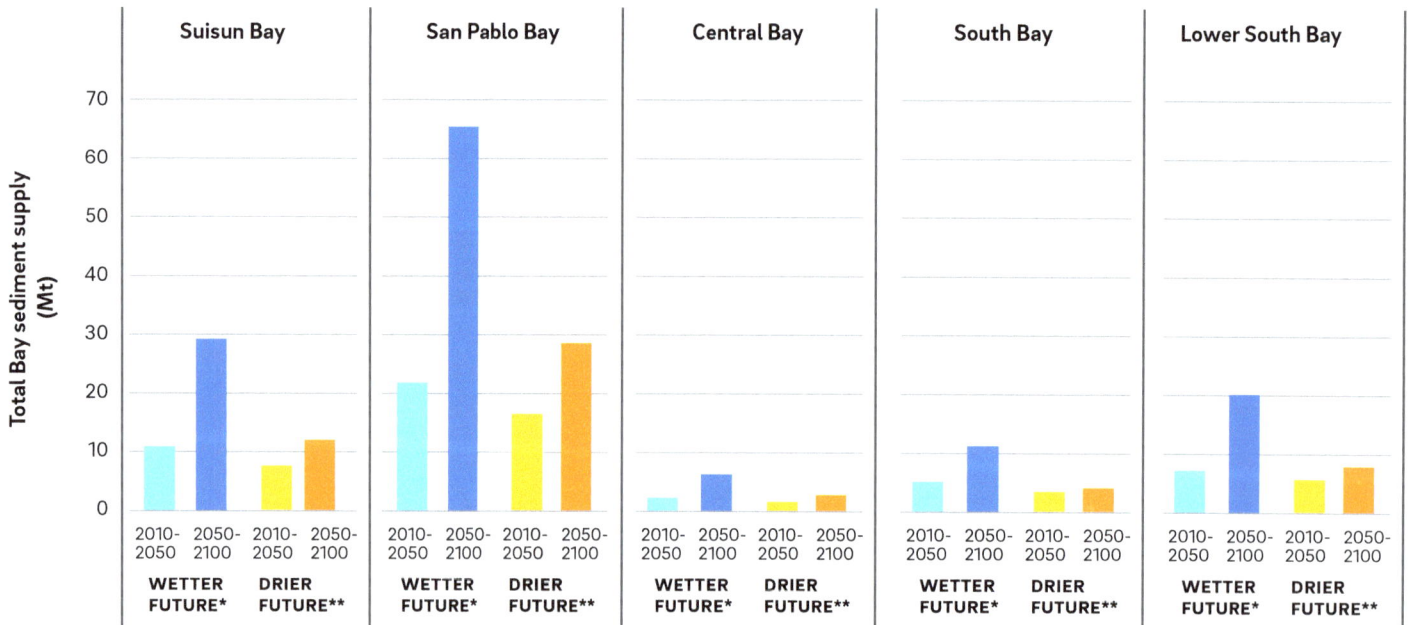

* Source for Wetter Future: CESM1-BGC 8.5   **Source for Drier Future: HadGEM2-CC 8.5

century. For the WY2010–2050 time period, the wetter and drier futures have modestly different total sediment supplies (80 Mt compared to 65 Mt), with the Delta providing a similar load for both futures (33 Mt or 41% of the total for the wetter future compared to 30 Mt or 46% of the total for the drier future). For WY2050–2100, however, the wetter future has a total sediment supply that is about twice that of the drier future (198 Mt compared to 94 Mt), with the Delta supplying 34% of the total for the wetter future and 42% of the total for the drier future.

Subembayment (Bay Tributaries)

The tributary sediment supply for the wetter and drier futures varies considerably by subembayment, with San Pablo Bay supply being the highest at ~50% of the total baywide tributary supply for both time periods for both futures and Central Bay tributary supply being the lowest at ~5% (Figure 2.17). Suisun Bay and San Pablo Bay tributaries combined contribute more than two times more sediment than Central Bay, South Bay, and Lower South Bay tributaries combined for each time period for both futures. For comparison, in the recent past, San Pablo Bay accounted for ~40% of the total tributary supply to the Bay, Central Bay accounted for ~4%, and Suisun Bay and San Pablo Bay tributaries combined contributed a little less than two times more than the tributaries from the other subembayments combined (Schoellhamer et al. 2018). For the WY2010–2050 time period, the wetter future supply is between 30% and 45% higher than the drier future, with Suisun Bay having the greatest difference between wetter and drier future supply and San Pablo Bay having the lowest difference. For WY 2050–2100, the wetter future supply is between 130% and 170% higher than the drier future, with South Bay having the greatest difference between wetter and drier future supply and Suisun Bay having the lowest difference.

San Pablo Bay, near Highway 37 and Sears Point Road (Photo by Shira Bezalel, SFEI)

SENSITIVITY ANALYSIS:
How sediment rating curve changes affect future sediment supply estimates

In this study, the future tributary sediment supply to the study area is driven by the sediment rating curves developed for historical conditions in the focus tributaries (WY1960–2016). As previously noted, the future relationship between annual runoff and annual sediment load may be quite different than the recent past due to a variety of factors, including changes to land use and precipitation intensity that affect both flow and sediment delivery. To assess how sensitive future sediment tributary load estimates can be to rating curves, we evaluated the relationship between changes in the curve and changes in load estimates for the Napa River watershed. For this analysis, we developed a suite of modified rating curves by holding the y-intercept of the Napa River curve constant and increasing the curve slope by 5%, 10%, and 25%, and then decreasing the curve slope by 5%, 10%, and 25% (Figure 2.18). We then used the suite of rating curves for the wetter and drier futures to calculate new estimates of Napa River sediment loads for 2010–2100 (Table 2.5). Increasing the rating curve slope by 5% led to total sediment load increasing by 95%, and increasing the rating curve slope by a factor of 25% led to total sediment load increasing by 2800%. Conversely, decreasing the rating curve slope by 5% led to total sediment load decreasing by 47% and decreasing the rating curve slope by 25% led to total sediment load decreasing by 93%. The dramatic difference at the higher slope increase/decrease factors is driven by the fact that the Napa River rating curve is a power function. This is not the case for all focus tributaries. Nonetheless, this analysis shows that total sediment load estimates can be quite sensitive to even modest rating curve changes. §

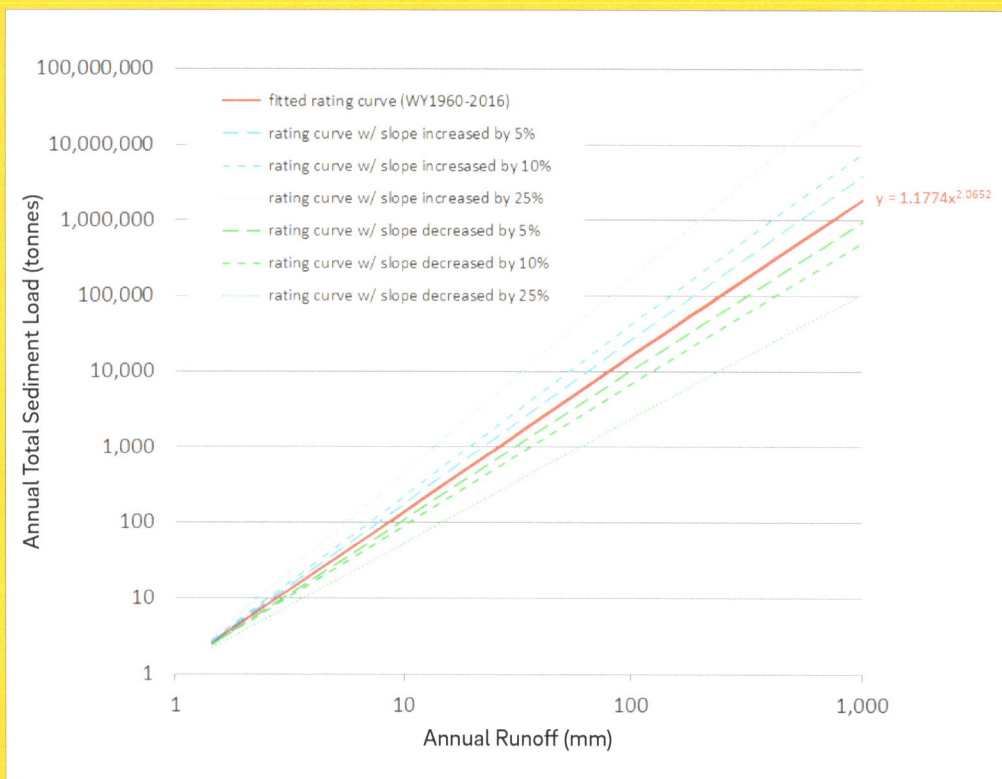

Figure 2.18. Napa River watershed rating curve (WY1960-2016) and suite of rating curves with increased and decreased slope used for sensitivity analysis.

Napa River at Vallejo (Courtesy of CC 2.0, photo by Zug Zwang)

Table 2.5. Results from Napa River rating curve sensitivity analysis.

Rating curve slope change amount	Total sediment load (WY2010–2100) change amount for both wetter and drier futures
5% increase	95% increase
10% increase	280% increase
25% increase	2800% increase
5% decrease	47% decrease
10% decrease	70% decrease
25% decrease	93% decrease

Napa River near Bay (Photo by WineCountry Media, courtesy CC 2.0)

Bayland Sediment Demand, Bay Sediment Supply, and Resilience

Introduction

The findings from the bayland sediment demand and Bay sediment supply analyses conducted for this study enable an understanding of future demand/supply ratios at a variety of spatial and temporal scales. When combined with organic matter accumulation rates, the findings can indicate the baylands that have the potential to accumulate inorganic and organic material needed to maintain their elevation as sea level continues to rise. In this chapter, we provide an overview of future bayland sediment demand compared to Bay sediment supply, extending out to the end of the century at different spatial scales. We then synthesize these findings with an assessment of organic matter accumulation rates into a map highlighting areas of high potential for long-term bayland resilience with respect to vertical accretion.

Comparison of Bayland Sediment Demand and Bay Sediment Supply

When assessing bayland sediment demand/supply ratios, it is important to recognize the uncertainty in the amount of sediment supply that will deposit onto tidal flats and tidal marshes. We know that not all of the sediment delivered to the Bay from the Delta and Bay tributaries will make it onto the baylands, and that the proportion that does reach the baylands depends on many factors such as channel alignment, stream power, the quality of the creek-to-bayland connection, and many other considerations. A large portion of this sediment deposits in the deep Bay in shipping channels, in ports and harbors, and in flood control channels downstream from the head of tide. Some of this sediment is stored only temporarily as large storms can scour and redistribute sediment from the Bay shallows onto tidal flats and tidal marshes. In addition, sediment available for bayland deposition can be transported out of the Bay through the Golden Gate. Although the amount of sediment that gets deposited on the baylands cannot be determined except in hindsight, we assume that it will be less than the sediment supply from the Delta and Bay tributaries. Figure 3.1 shows many of the factors, or groups of factors, that influence comparison of near-term sediment supply estimates (2010–2050) for the wetter future with near-term sediment demand estimates (1.9 ft (~0.6 m) of SLR) for both bayland restoration scenarios.

San Francisco Bay (Landsat imagery courtesy NASA, April 2013)

Figure 3.1. Conceptual understanding of the relationship between total Bay sediment supply and the proportion available for bayland deposition. Based on near-term sediment supply estimates for the wetter future and near-term baylands sediment demand estimates for both bayland habitat scenarios.

Tributary sediment supply at head of tide: The total net tributary sediment supply into SF Bay between 2010 and 2050 under a wetter climate future (CESM1-BGC 8.5) is approximately 80 Mt. About 40% of this estimate is based on sediment from Delta tributaries and 60% from Bay tributaries.

Sediment transport downstream of head of tide: As sediment travels out of the watershed and into tidal reaches, sediment supply to the baylands decreases as a result of in-channel sedimentation and deposition in the Bay's deep channels. Additional sediment will likely be lost from flux through the Golden Gate to the Pacific Ocean, the magnitude of which is unknown.

Sediment deposition onto baylands habitats: Existing bayland habitats (ca. 2009) will need approximately 105 Mt of sediment to keep pace with 1.9 ft of SLR by 2050. If all tidal marsh restoration projects are underway during this time period, an additional 119 Mt of sediment will be needed to meet bayland habitat demands. More data is needed to determine the amount of sediment from Bay and Delta tributaries that would be deposited onto bayland habitats.

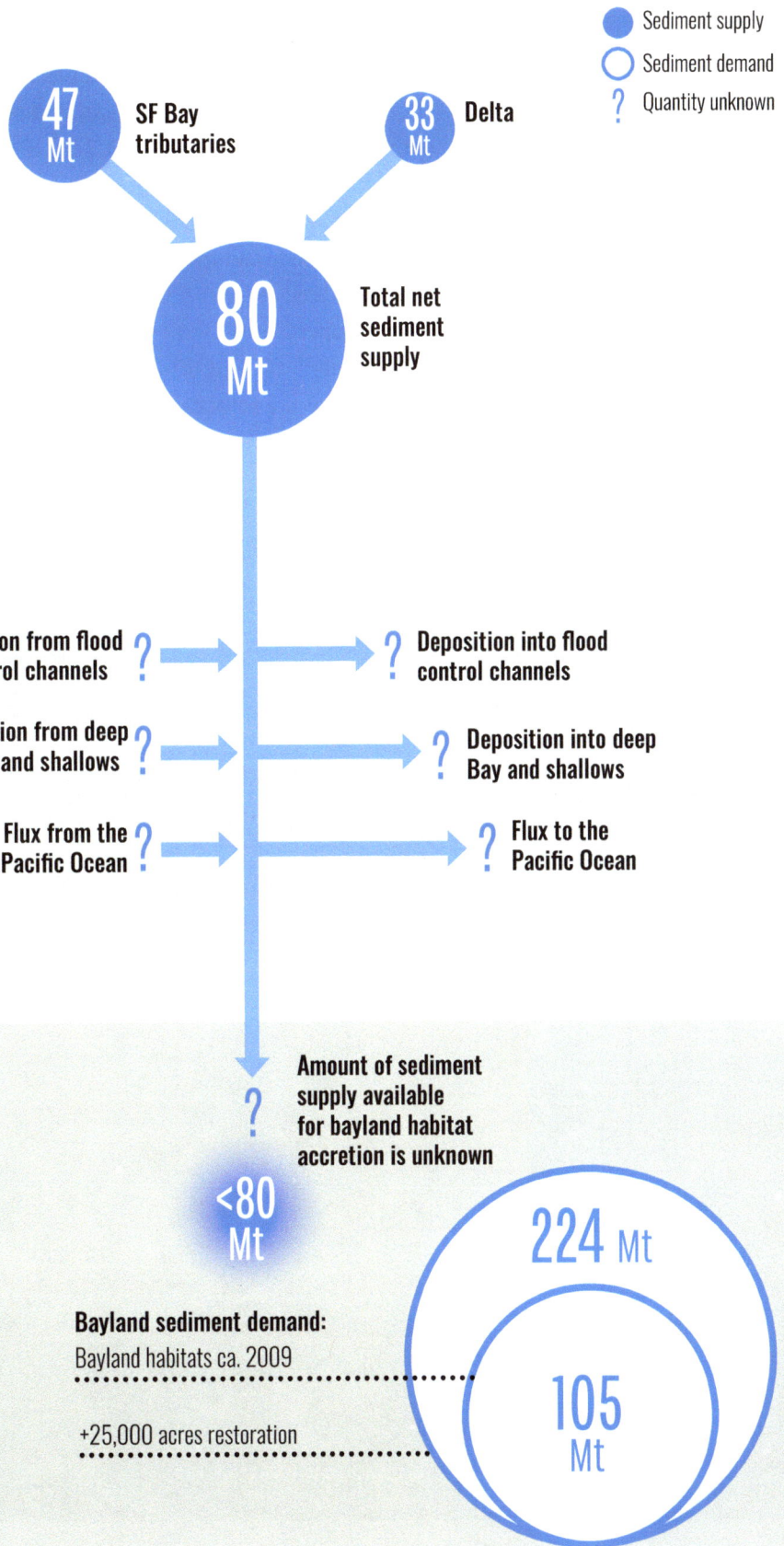

47 Mt SF Bay tributaries

33 Mt Delta

80 Mt Total net sediment supply

● Sediment supply
○ Sediment demand
? Quantity unknown

Erosion from flood control channels **?** → **?** Deposition into flood control channels

Erosion from deep Bay and shallows **?** → **?** Deposition into deep Bay and shallows

Flux from the Pacific Ocean **?** → **?** Flux to the Pacific Ocean

? Amount of sediment supply available for bayland habitat accretion is unknown

<80 Mt

Bayland sediment demand:
Bayland habitats ca. 2009 ....

+25,000 acres restoration ....

224 Mt

105 Mt

Full Bay

For both the wetter and drier futures, our analysis shows that the projected sediment supply to the Bay from the Delta and Bay tributaries is not enough to supply all existing and planned restored baylands with the sediment needed to keep pace with SLR until the end of the 21st century (Figure 3.2). The calculated amount of sediment needed to maintain the elevation of existing baylands for 1.9 ft (~0.6 m) of SLR and an additional 5.0 ft (~1.5 m) of SLR is of the same general magnitude as the calculated amount of sediment supplied to the Bay in the wetter future for the 2010–2050 and 2050–2100 time periods, but only a portion of the sediment supply would likely remain in the Bay and be available for deposition on baylands (as discussed above). When considering the amount of sediment needed to support both existing baylands + planned restoration, the wetter future supply is approximately 50% of total bayland demand and the drier future supply is only approximately 30% of the total demand (again, with only a portion of that supply actually available for deposition on baylands).

Central San Francisco Bay (Photo by Shira Bezalel, SFEI)

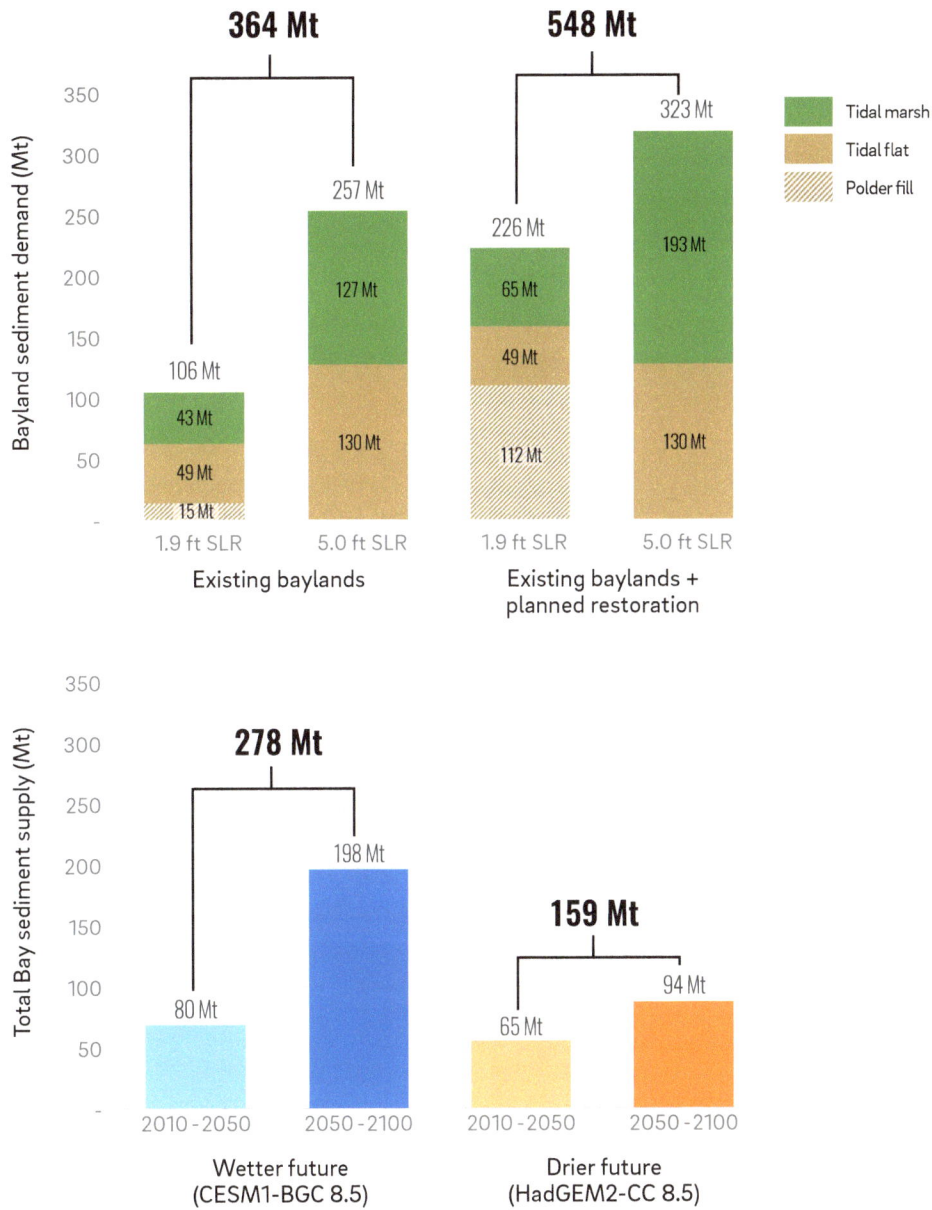

Figure 3.2. Comparison of total bayland sediment demand and total Bay sediment supply for the wetter future (CESM1-BGC 8.5) and drier future (HadGEM2-CC 8.5) for WY2010–2050 (1.9 ft (0.6 m) SLR) and WY2050-2100 (5.0 ft (1.5 m) SLR).

Subregional Scale

Suisun Bay–San Pablo Bay

Our analysis shows that the Delta and Bay tributary sediment supply for both the wetter and drier futures is similar to the amount needed to support just existing baylands between 2010 and 2050, but only the wetter future has a sediment supply high enough to support both existing and planned restored baylands between 2050 and 2100. For 1.9 ft (0.6 m) of SLR between 2010 and 2050, the bayland sediment demand is 57 Mt for existing baylands and 119 Mt for existing baylands + planned restoration (with polder filling accounting for almost half of the demand) (Figure 3.3a). During this time period, the wetter future sediment supply is 66 Mt and the drier future sediment supply is lower at 54 Mt. For 5.0 ft (1.5 m) of SLR between 2050 and 2100, the bayland sediment demand is 128 Mt for existing baylands and 156 Mt for existing baylands and restored marshes (Figure 3.3b). The wetter future sediment supply during this time period is 161 Mt but the drier future sediment supply is only 80 Mt, which is approximately 60% of the sediment needed for just the existing baylands. It is important to note that future net sediment supply to Suisun Bay–San Pablo Bay is thought to be less than the Delta and Bay tributary sediment supply due to the assumed future annual net flux towards Central Bay.

Central Bay

The findings for Central Bay show a situation similar to Suisun Bay–San Pablo Bay. Sediment supply for both the wetter and drier futures is similar to the amount needed to support just existing baylands between 2010 and 2050, but the wetter future has a sediment supply high enough to support both existing and planned restored baylands between 2050 and 2100. For 1.9 ft (0.6 m) of SLR between 2010 and 2050, the bayland sediment demand is 8 Mt for both bayland scenarios because the acreage of planned restored marsh is very low (approximately 100 acres (~40 ha), which is only a 3% increase above current marsh acreage) (Figure 3.3a). During this time period, the wetter future and drier future sediment supply from Bay tributaries is 2 Mt, but the total sediment supply is assumed to be similar to if not greater than bayland sediment demand when considering sediment influx from San Pablo Bay and South Bay and outflux at the Golden Gate. For 5.0 ft (1.5 m) of SLR between 2050 and 2100, the sediment demand for both bayland scenarios is 22 Mt (Figure 3.3b). The wetter future tributary sediment supply during this time period is 6 Mt and the drier future tributary sediment supply is 3 Mt. When accounting for San Pablo Bay, South Bay, and Golden Gate fluxes, the wetter future may have a total sediment supply similar to bayland sediment demand.

South Bay–Lower South Bay

Unlike the other subregions, the findings for South Bay–Lower South Bay suggest that bayland sediment demand is much greater than sediment supply for both time periods and both bayland scenarios in both the wetter and drier future (Figure 3.3). For 1.9 ft (0.6 m) of SLR between 2010 and 2050, the bayland sediment demand is 40 Mt for existing baylands and 98 Mt for existing baylands + planned restoration (with polder filling accounting for almost half of the demand) (Figure 3.3a). During this time period, wetter future sediment supply is 12 Mt (~30% of the existing baylands demand) and the drier future sediment supply is 9 Mt (~20% of the existing baylands demand). For 5.0 ft (1.5 m) of SLR between 2050 and 2100, the bayland sediment demand is 108 Mt for existing baylands and 145 Mt for existing baylands and restored marshes (Figure 3.3b). The wetter future sediment supply during this time period is 30 Mt (~30% of the existing baylands demand) and the drier future sediment supply is 12 Mt (~10% of the existing baylands demand). When accounting for the assumed annual net sediment flux towards Central Bay, the difference between bayland sediment demand and sediment supply grows even greater.

Near-term future
2010 - 2050, 1.9 ft SLR

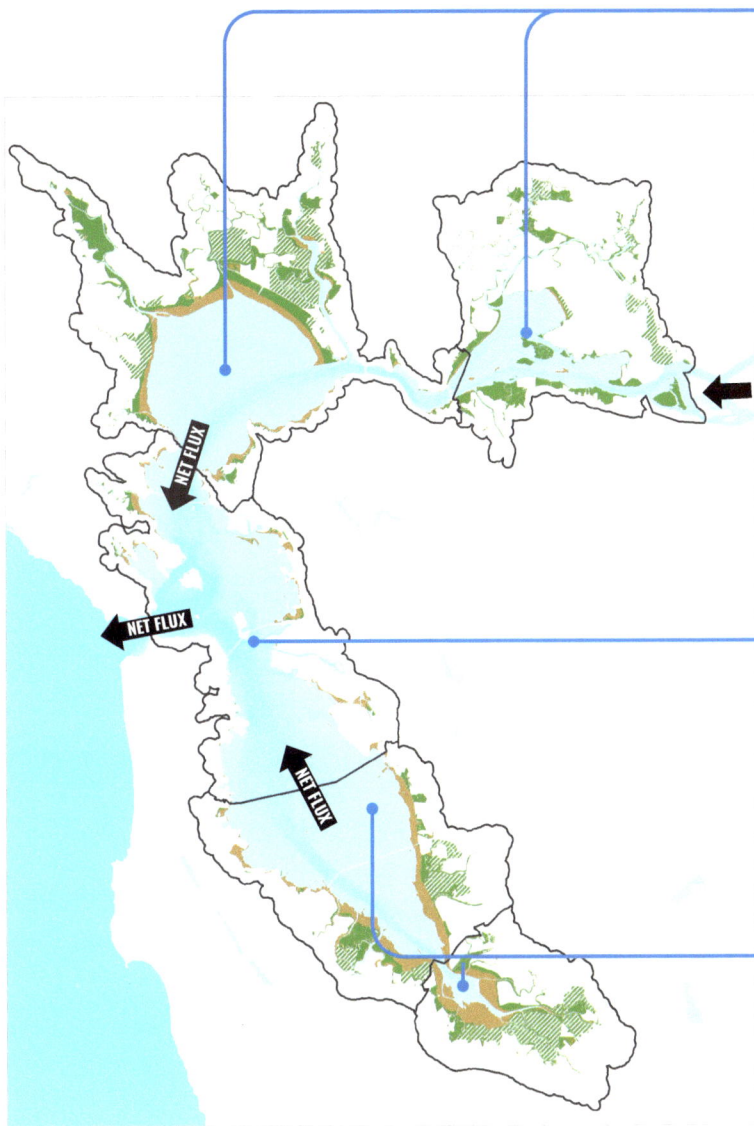

Figure 3.3a. Subregional bayland sediment demand and Bay sediment supply for existing baylands and existing baylands plus restoration for 2010-2050 and for the wetter future (CESM1-BGC 8.5) and the drier future (HadGEM2-CC 8.5).

Map legend

—— Subembayment boundary

Existing baylands

Tidal flat

Tidal marsh

Planned and in-progress restoration

/// Tidal marsh

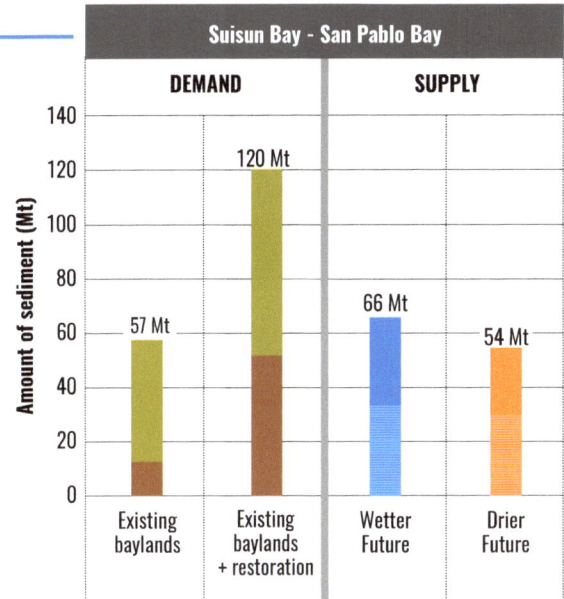

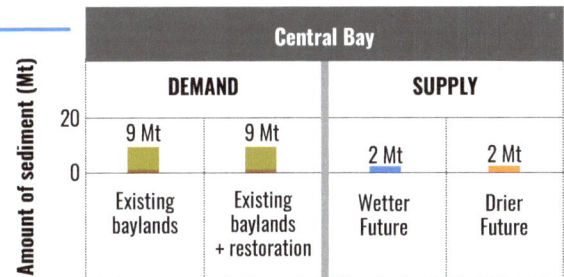

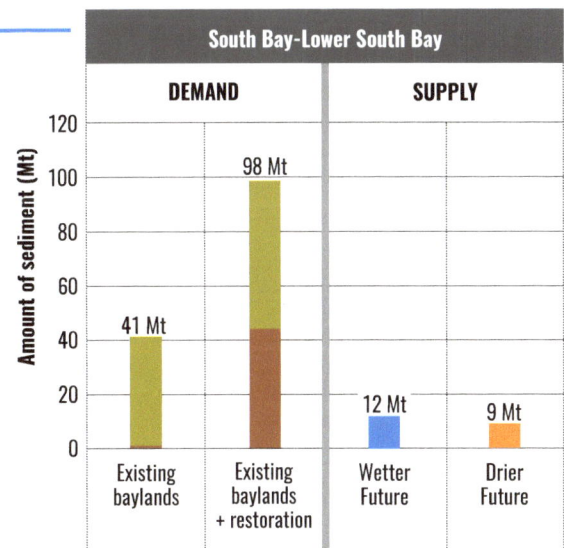

Bar chart legend

NATURAL DEPOSITION

POLDER FILL

WETTER | DRIER

PORTION FROM LOCAL TRIBUTARIES

PORTION FROM DELTA

Suisun Bay - San Pablo Bay

| DEMAND | | SUPPLY | |

- Existing baylands: 57 Mt
- Existing baylands + restoration: 120 Mt
- Wetter Future: 66 Mt
- Drier Future: 54 Mt

Central Bay

| DEMAND | | SUPPLY | |

- Existing baylands: 9 Mt
- Existing baylands + restoration: 9 Mt
- Wetter Future: 2 Mt
- Drier Future: 2 Mt

South Bay-Lower South Bay

| DEMAND | | SUPPLY | |

- Existing baylands: 41 Mt
- Existing baylands + restoration: 98 Mt
- Wetter Future: 12 Mt
- Drier Future: 9 Mt

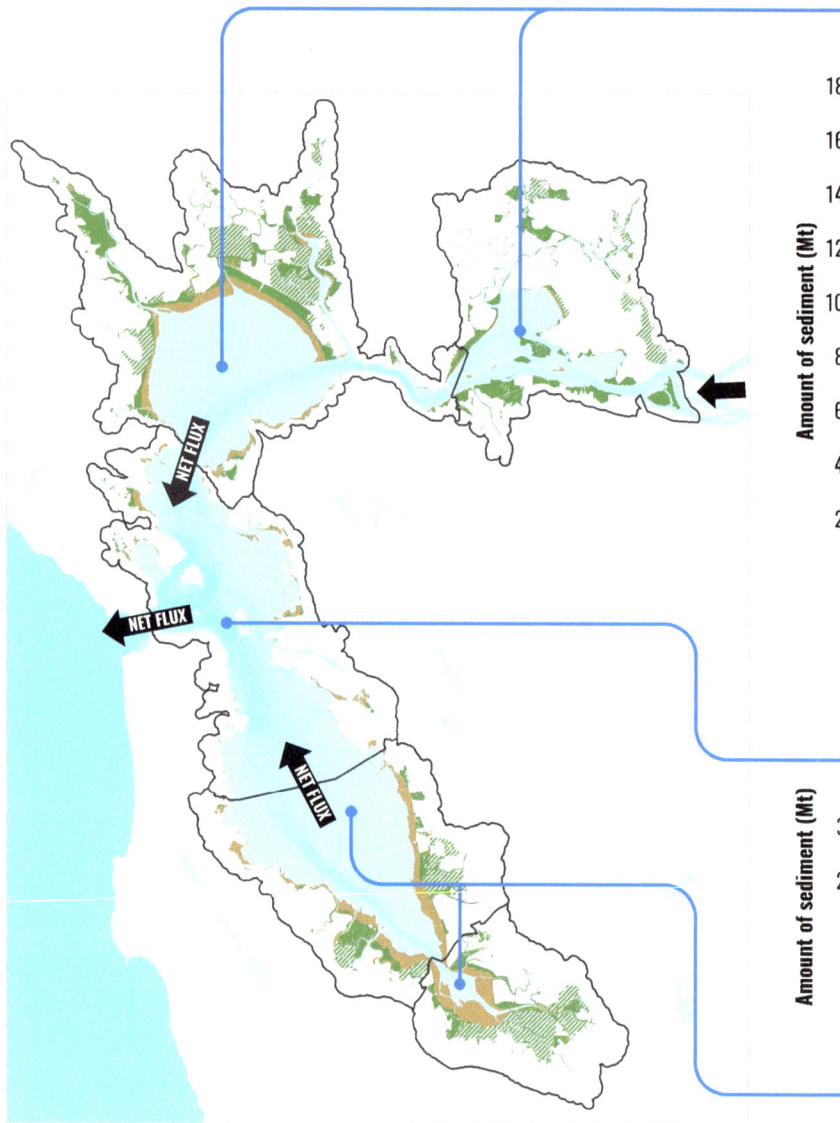

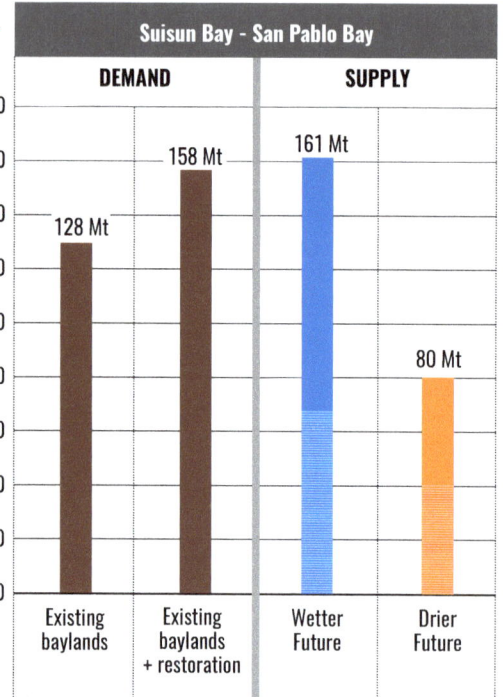

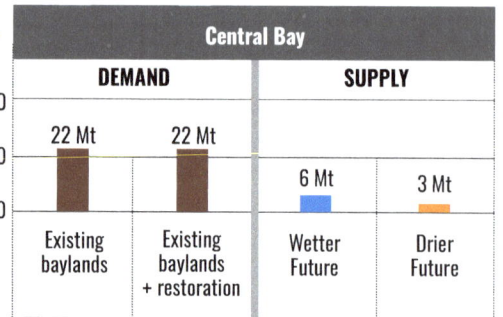

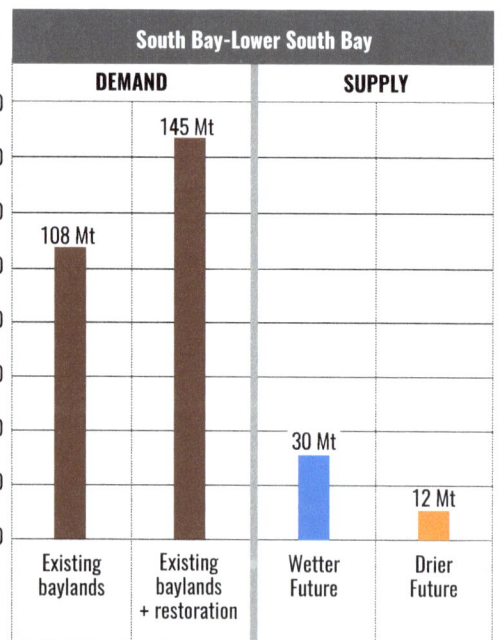

Figure 3.3b. Subregional bayland sediment demand and Bay sediment supply for existing baylands and existing baylands plus restoration for 2050-2100 for the wetter future (CESM1-BGC 8.5) and the drier future (HadGEM2-CC 8.5).

Map legend

— Subembayment boundary

Existing baylands

■ Tidal flat

■ Tidal marsh

Planned and in-progress restoration

▨ Tidal marsh

Bar chart legend

■ NATURAL DEPOSITION

WETTER / DRIER

■■ PORTION FROM LOCAL TRIBUTARIES

▨▨ PORTION FROM DELTA

Suisun Bay - San Pablo Bay

| DEMAND | | SUPPLY | |

Amount of sediment (Mt)

- Existing baylands: 128 Mt
- Existing baylands + restoration: 158 Mt
- Wetter Future: 161 Mt
- Drier Future: 80 Mt

Central Bay

| DEMAND | | SUPPLY | |

Amount of sediment (Mt)

- Existing baylands: 22 Mt
- Existing baylands + restoration: 22 Mt
- Wetter Future: 6 Mt
- Drier Future: 3 Mt

South Bay-Lower South Bay

| DEMAND | | SUPPLY | |

Amount of sediment (Mt)

- Existing baylands: 108 Mt
- Existing baylands + restoration: 145 Mt
- Wetter Future: 30 Mt
- Drier Future: 12 Mt

OLU Scale

Assessing the ratio of bayland sediment demand to tributary sediment supply at the OLU scale is useful for understanding how far local tributary sediment could go towards supporting adjacent baylands. For OLUs with polders that are slated for restoration, assessing the ratio with and without polder filling before restoration provides an indication of how much the polders are driving bayland demand. OLUs with low demand to supply ratios for both the wetter and drier futures are the areas where sediment demand could be addressed in large part by local tributary sediment supply, given appropriate management measures. Below, we show the bayland sediment demand to tributary sediment supply ratios for existing and restored baylands for both time periods in the wetter and drier future for all OLUs. For OLUs with polders, we also show the ratios without and with polder fill before restoration for the 2010–2050 time period.

Walnut, Carquinez South, Suisun Slough, and Richardson Bay OLUs have bayland sediment demand that is similar to or less than local tributary sediment supply for both time periods for both the wetter and drier futures (Figure 3.4a and b). Pinole, Petaluma, Wildcat, and Colma–San Bruno OLUs all have relatively low bayland demand to local tributary sediment supply ratios (between two and five) for both time periods for the wetter and drier futures. These eight OLUs are considered to have the greatest potential for bayland sediment demand to be addressed in large part by local tributary sediment supply. Several other OLUs have relatively low ratios for the wetter future but bayland sediment demand that is much higher than tributary sediment supply for the drier future, particularly during the 2050–2100 time period. The OLUs with ratios that decrease for the 2010–2050 time period if polders are filled before restoration include Napa-Sonoma, Montezuma Slough, Novato, Alameda, Stevens, Santa Clara Valley, and Belmont-Redwood. The magnitude of decrease in the ratio ranges from 20% for Alameda (a ratio of 5 without polder fill and 4 with polder fill for the wetter future) to -80% for Montezuma Slough (a ratio of 22 without polder fill and 4 with polder fill for the drier future).

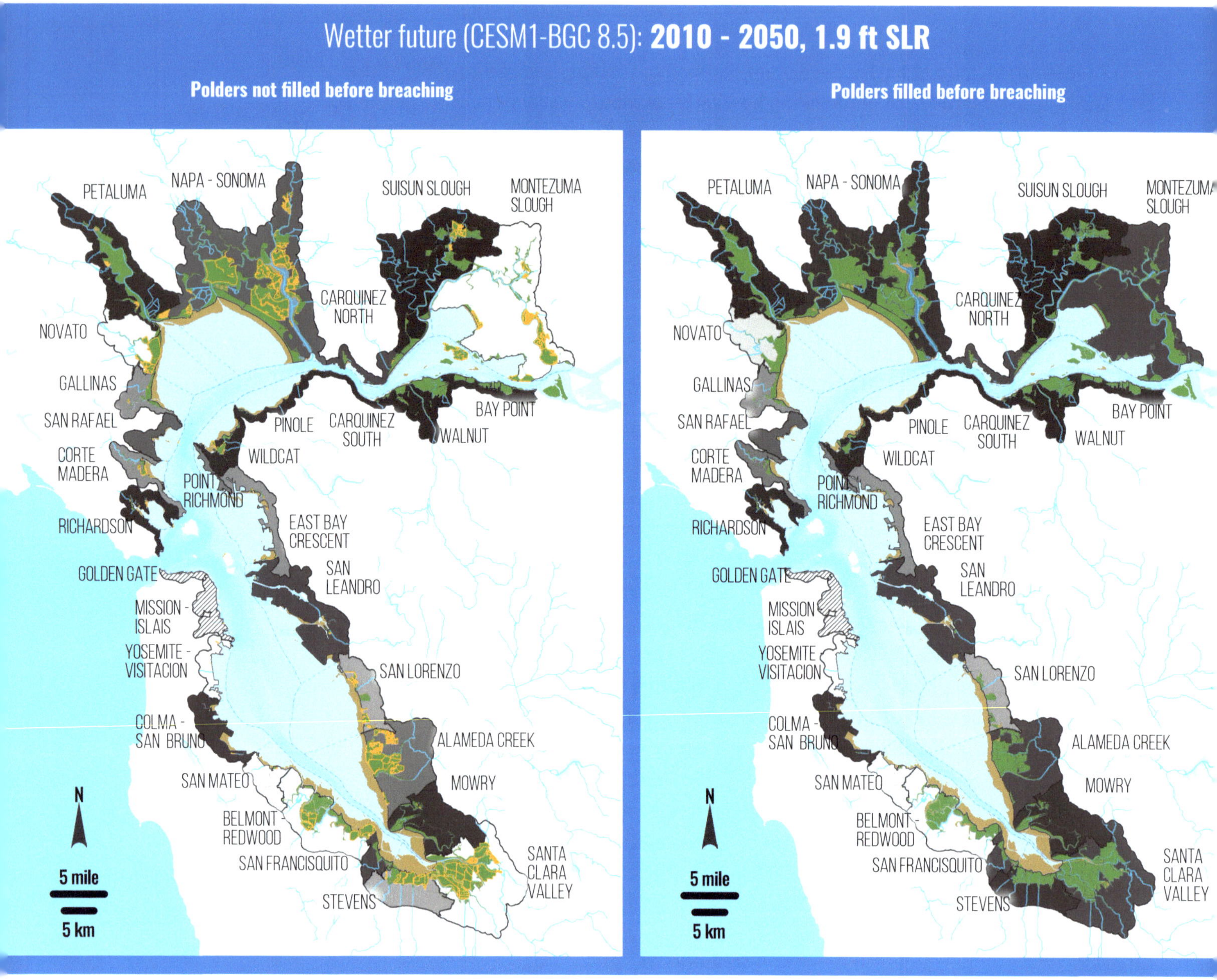

Wetter future (CESM1-BGC 8.5): **2010 - 2050, 1.9 ft SLR**
Polders not filled before breaching
Polders filled before breaching
PETALUMA
NAPA - SONOMA
SUISUN SLOUGH
MONTEZUMA SLOUGH
CARQUINEZ NORTH
NOVATO
GALLINAS
SAN RAFAEL
PINOLE
CARQUINEZ SOUTH
BAY POINT
WALNUT
CORTE MADERA
WILDCAT
POINT RICHMOND
RICHARDSON
EAST BAY CRESCENT
GOLDEN GATE
SAN LEANDRO
MISSION - ISLAIS
YOSEMITE - VISITACION
SAN LORENZO
COLMA - SAN BRUNO
ALAMEDA CREEK
SAN MATEO
MOWRY
BELMONT - REDWOOD
SAN FRANCISQUITO
SANTA CLARA VALLEY
STEVENS
N
5 mile
5 km

Figure 3.4a. Ratio of bayland sediment demand to local tributary sediment supply for existing baylands + planned restoration for 2010–2050 (with and without filling polders before restoration) and 2050–2100 for the wetter future (CESM1-BGC 8.5).

Legend FOR THE MAPS ON THIS AND FOLLOWING PAGE

Creek

EXISTING & PLANNED HABITATS

Mudflat

Tidal marsh

OTHER CONSIDERATIONS

Polders

Insufficient data

RATIO OF LOCAL BAYLANDS SEDIMENT DEMAND TO LOCAL TRIBUTARY SEDIMENT SUPPLY

<1: Supply exceeds demand

>10: Demand 10x greater than supply

Wetter future (CESM1-BGC 8.5): 2050 - 2100, 5.0 ft SLR

Assumes all polders have been converted to tidal marsh, and that all tidal flats and marshes have kept pace with SLR by 2050

PETALUMA

NAPA - SONOMA

SUISUN SLOUGH

MONTEZUMA SLOUGH

CARQUINEZ NORTH

NOVATO

GALLINAS

BAY POINT

SAN RAFAEL

PINOLE

CARQUINEZ SOUTH

WALNUT

CORTE MADERA

WILDCAT

POINT RICHMOND

RICHARDSON

EAST BAY CRESCENT

GOLDEN GATE

SAN LEANDRO

MISSION ISLAIS

YOSEMITE VISITACION

SAN LORENZO

COLMA - SAN BRUNO

ALAMEDA CREEK

SAN MATEO

MOWRY

BELMONT REDWOOD

SAN FRANCISQUITO

SANTA CLARA VALLEY

STEVENS

N

5 mile

5 km

Polders not filled before breaching

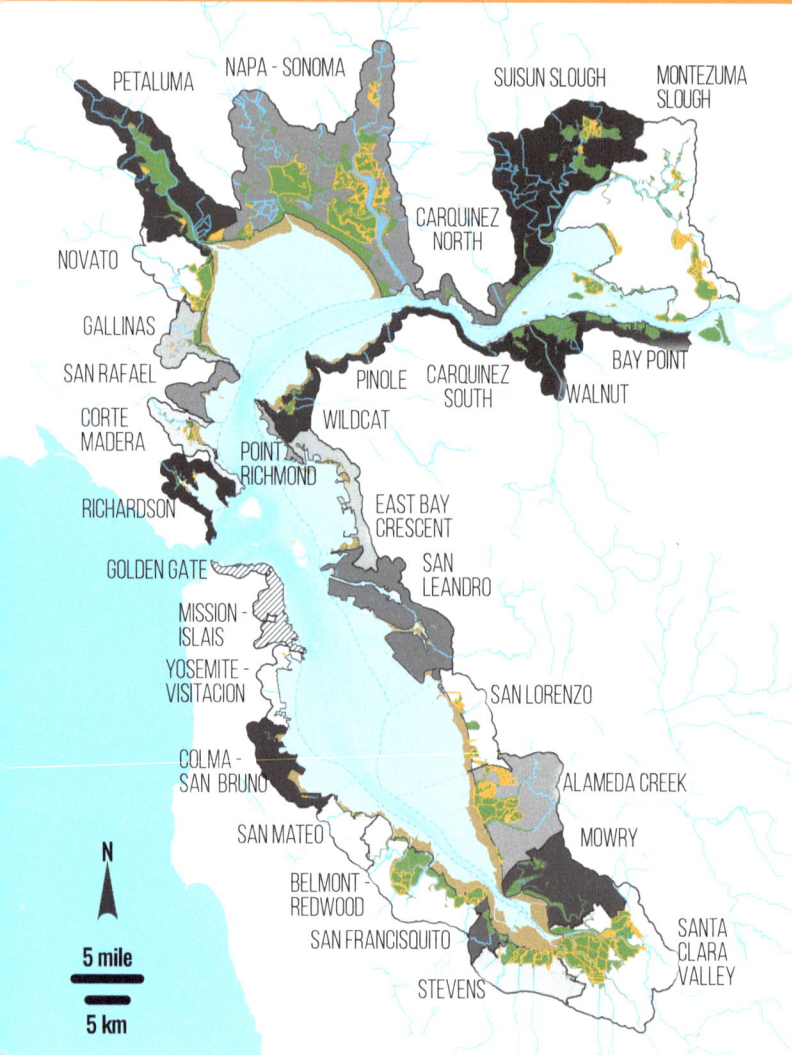

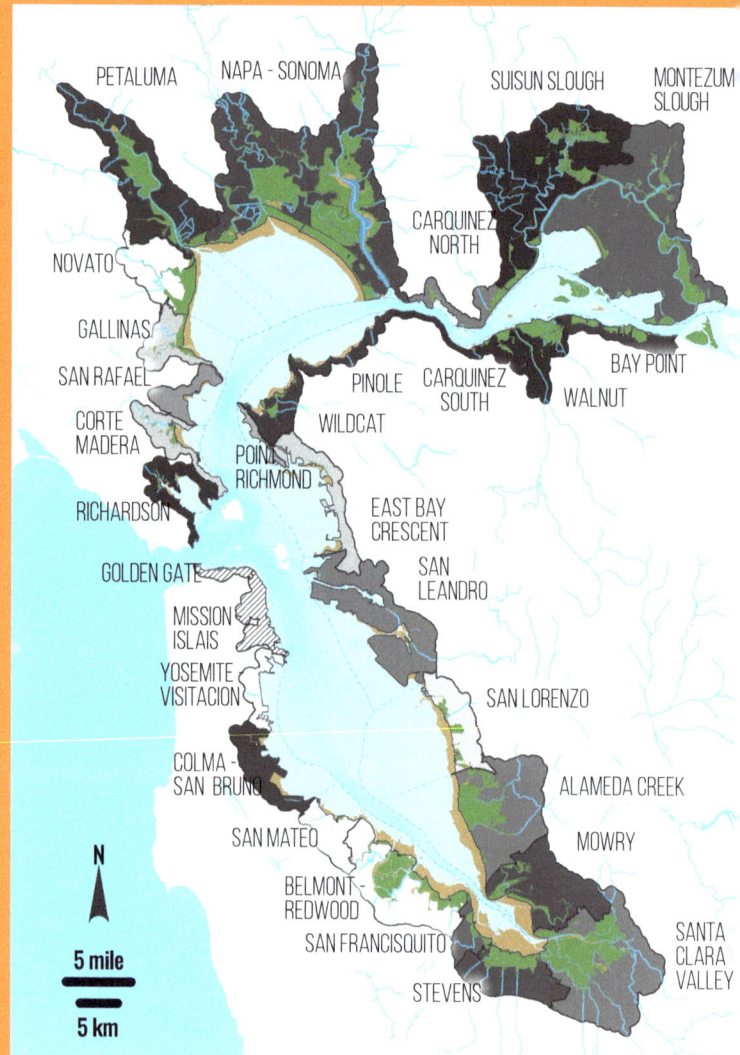

Polders filled before breaching

Figure 3.4b. Ratio of bayland sediment demand to local tributary sediment supply for existing baylands + planned restoration for 2010–2050 (with and without filling polders before restoration) and 2050–2100 and the drier future (HadGEM2-CC 8.5).

Legend FOR THE MAPS ON THIS AND FOLLOWING PAGE

— Creek

EXISTING & PLANNED HABITATS

Mudflat

Tidal marsh

OTHER CONSIDERATIONS

Polders

Insufficient data

RATIO OF LOCAL BAYLANDS SEDIMENT DEMAND TO LOCAL TRIBUTARY SEDIMENT SUPPLY

<1: Supply exceeds demand

>10: Demand 10x greater than supply

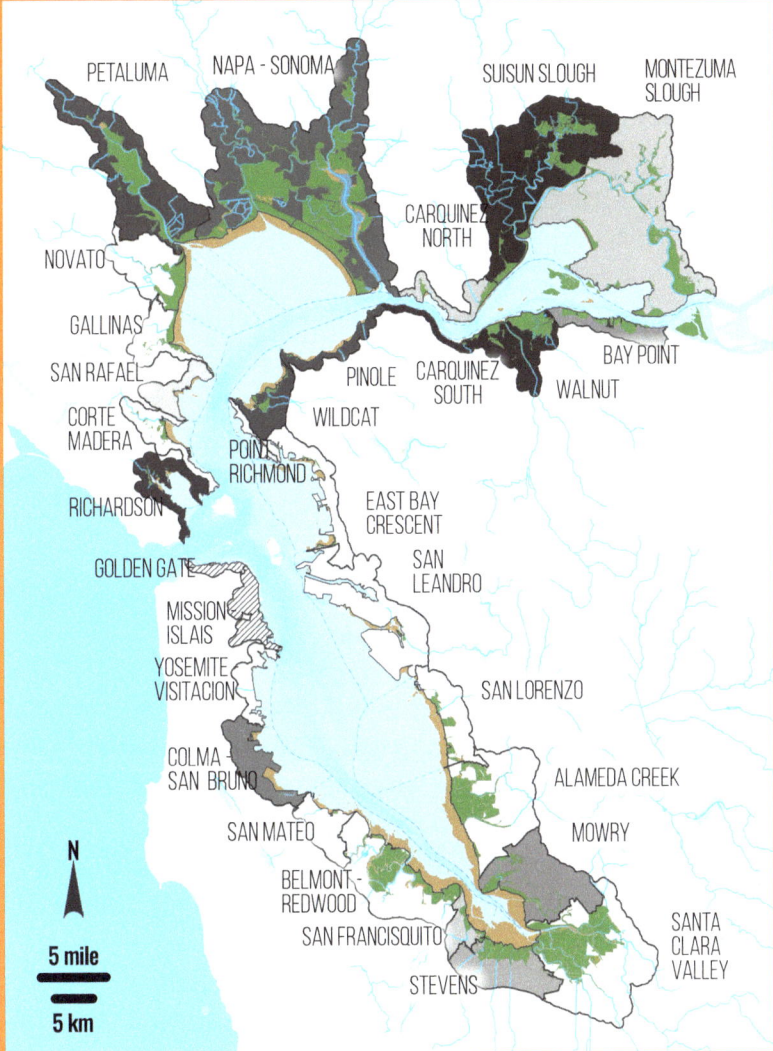

Drier future (HadGEM2-CC 8.5): **2050 - 2100, 5.0 ft SLR**

Assumes all polders have been converted to tidal marsh, and that all tidal flats and marshes have kept pace with SLR by 2050

PETALUMA

NAPA - SONOMA

SUISUN SLOUGH

MONTEZUMA SLOUGH

CARQUINEZ NORTH

NOVATO

GALLINAS

SAN RAFAEL

CORTE MADERA

RICHARDSON

GOLDEN GATE

PINOLE

WILDCAT

POINT RICHMOND

CARQUINEZ SOUTH

BAY POINT

WALNUT

EAST BAY CRESCENT

SAN LEANDRO

MISSION ISLAIS

YOSEMITE VISITACION

COLMA - SAN BRUNO

SAN MATEO

BELMONT - REDWOOD

SAN FRANCISQUITO

STEVENS

SAN LORENZO

ALAMEDA CREEK

MOWRY

SANTA CLARA VALLEY

N

5 mile

5 km

Organic Matter Accumulation

In addition to mineral sediment supply, organic matter accumulation can be a key factor in helping tidal marsh accretion rates keep pace with a rising sea level. Typically, brackish to freshwater marshes are dominated by highly productive vegetation that contributes to higher rates of organic matter accumulation than salt marsh vegetation (Stralberg et al. 2011). Tidal marshes with high organic matter production rates have been shown in some instances to accrete faster than those predominantly composed of inorganic sediment (Callaway et al. 1996, Morris et al. 2002). Organic matter accumulation has also been shown to outpace inorganic sediment accumulation, accounting for more elevation per mass unit under conditions of limited inorganic sediment supply (Swanson et al. 2014). Within the Bay-Delta ecosystem, the oligohaline marshes of Suisun Bay have been shown to have higher organic matter accumulation rates than San Francisco Bay salt marshes (Callaway et al. 2012). In addition, in freshwater Delta marshes dominated by high productivity vegetation such as cattails (*Typha spp.*) and bulrushes (*Schoenoplectus spp.*), organic matter accumulation rates have been shown to be more than four times greater than the highest rate of organic matter accumulation measured in tidal marshes around San Francisco Bay (Deverel et al. 2008, Drexler et al. 2009, Swanson et al. 2014). The maximum rate of SLR that organic matter accumulation in the Bay can match, however, remains poorly understood.

Based on this previous research, the oligohaline marshes in Suisun Bay are here assumed to have a relatively high rate of organic matter accumulation, the brackish marshes in Suisun Bay and San Pablo Bay are assumed to have a relatively moderate rate of organic matter accumulation, and organic matter accumulation in all salt marshes is assumed to be minimal. There are localized zones of relatively high organic matter accumulation rates in some salt and brackish marsh around the Bay at head of tide in large tributaries and at treated wastewater discharge points (Goals Update 2015). Although these will be critical resilient marsh "nodes" as sea level continues to rise, they occur at a spatial scale that is too small to be considered in this study.

Assessment of Bayland Resilience

Combining the findings from this study with our qualitative assessment of organic matter accumulation rates provides an indication of the baylands with the highest potential for long-term resilience with respect to vertical accretion for both a wetter or drier future (Figure 3.5).

Suisun Bay–San Pablo Bay

Based on the findings from the future bayland sediment demand and region-wide sediment supply analyses, the Suisun Bay–San Pablo Bay subregion could lack the sediment supply needed to maintain the elevation of all existing and planned restored baylands to 2100. Even though the wetter future shows a bayland sediment demand similar to sediment supply for the later part of the 21st century, only a portion of that sediment supply would be available to the baylands.

When considering local bayland demand and the calculated supply of inorganic sediment plus assumed organic matter accumulation rates, several areas of high potential bayland resilience with respect to elevation become apparent. In Suisun Bay, the existing and planned restored marshes to the north and south could capture a large portion of the Delta's sediment supply, and the Delta's freshwater inflow will support lower salinity tidal marshes with relatively high organic matter production and accumulation rates. In addition, low local bayland demand to local tributary sediment supply ratios for both a wetter and drier future in the Suisun Slough, Walnut, Carquinez South OLUs indicate that local tributary supply could go a long way toward meeting bayland sediment demand if that sediment can be directed towards and deposited on local baylands. The oligohaline Montezuma Slough OLU baylands could have a high potential for bayland resilience with respect to vertical accretion if the polders are filled before restoration ***and*** tributary sediment is deposited onto the baylands. In San Pablo Bay, the Petaluma OLU baylands have relatively low local demand and local tributary sediment supply for both a wetter and drier future, indicating that local tributary supply could help meet future bayland demand with the right management approaches to trap tributary sediment onto baylands. The brackish Napa-Sonoma OLU baylands complex could also have a high potential for bayland resilience with respect to vertical accretion due to relatively high local sediment supply and organic matter accumulation rates, and because a portion of the sediment demand has been met for restoration projects implemented since 2010 (e.g., Cullinan Ranch).

Central Bay

Based on the findings from the future bayland sediment demand and region-wide sediment supply analyses, Central Bay could have the sediment supply needed to maintain the elevation of existing and planned restored baylands under a wetter future. A wetter future would have more sediment delivery from local tributaries and the Delta,

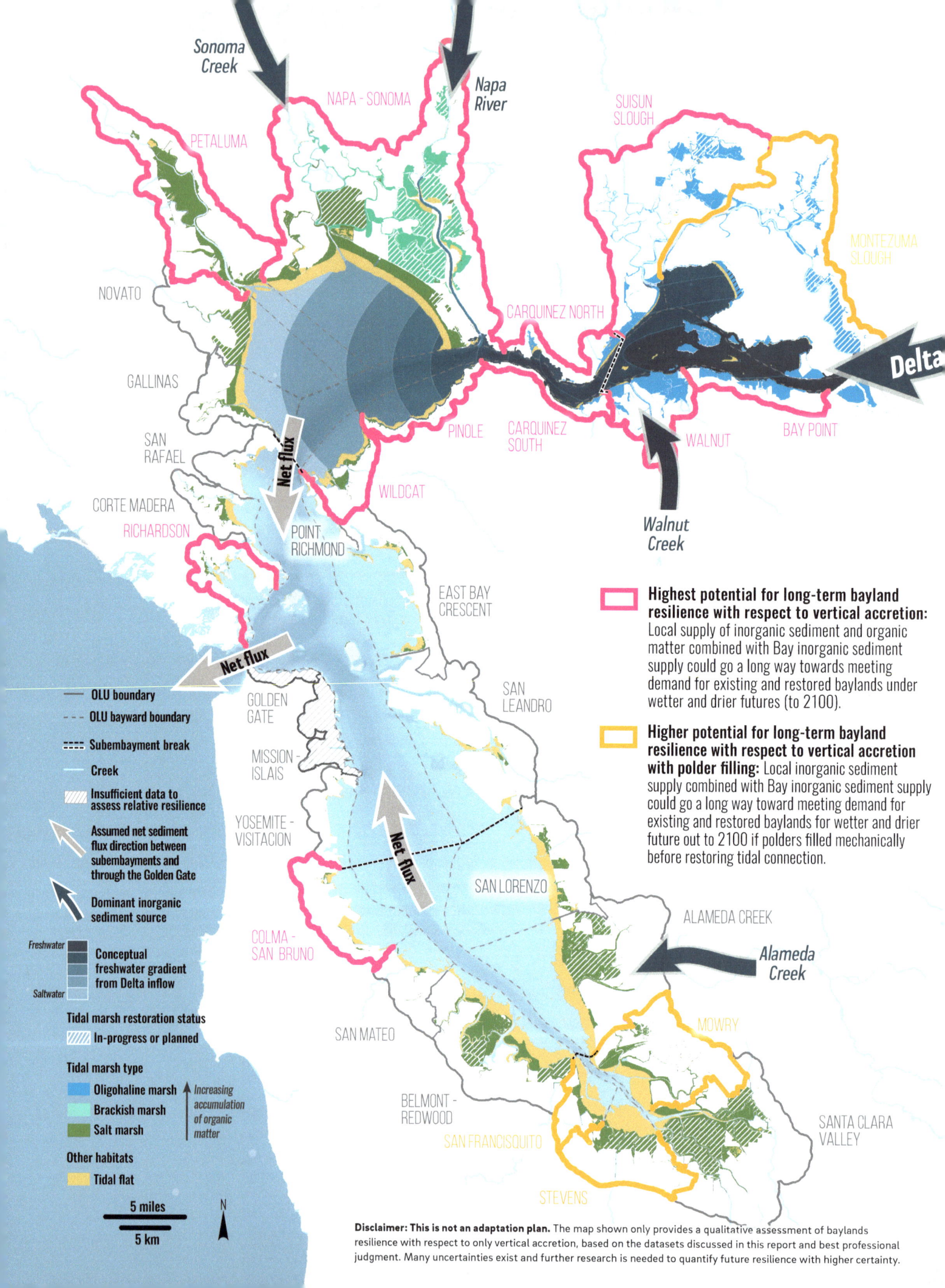

Sonoma Creek

Napa River

NAPA - SONOMA

SUISUN SLOUGH

PETALUMA

MONTEZUMA SLOUGH

NOVATO

CARQUINEZ NORTH

Delta

GALLINAS

SAN RAFAEL

PINOLE

CARQUINEZ SOUTH

WALNUT

BAY POINT

Net flux

CORTE MADERA

RICHARDSON

WILDCAT

POINT RICHMOND

Walnut Creek

EAST BAY CRESCENT

Net flux

SAN LEANDRO

GOLDEN GATE

MISSION - ISLAIS

YOSEMITE - VISITACION

Net flux

SAN LORENZO

ALAMEDA CREEK

COLMA - SAN BRUNO

Alameda Creek

SAN MATEO

MOWRY

BELMONT - REDWOOD

SANTA CLARA VALLEY

SAN FRANCISQUITO

STEVENS

Highest potential for long-term bayland resilience with respect to vertical accretion: Local supply of inorganic sediment and organic matter combined with Bay inorganic sediment supply could go a long way towards meeting demand for existing and restored baylands under wetter and drier futures (to 2100).

Higher potential for long-term bayland resilience with respect to vertical accretion with polder filling: Local inorganic sediment supply combined with Bay inorganic sediment supply could go a long way toward meeting demand for existing and restored baylands for wetter and drier future out to 2100 if polders filled mechanically before restoring tidal connection.

OLU boundary

OLU bayward boundary

Subembayment break

Creek

Insufficient data to assess relative resilience

Assumed net sediment flux direction between subembayments and through the Golden Gate

Dominant inorganic sediment source

Freshwater

Conceptual freshwater gradient from Delta inflow

Saltwater

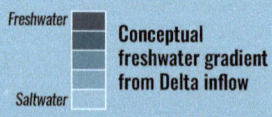

Tidal marsh restoration status
In-progress or planned

Tidal marsh type
Oligohaline marsh
Brackish marsh
Salt marsh

Increasing accumulation of organic matter

Other habitats
Tidal flat

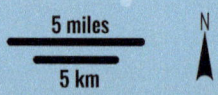

5 miles

5 km

N

Disclaimer: This is not an adaptation plan. The map shown only provides a qualitative assessment of baylands resilience with respect to only vertical accretion, based on the datasets discussed in this report and best professional judgment. Many uncertainties exist and further research is needed to quantify future resilience with higher certainty.

which would result in a large volume of sediment moving into Central Bay from both the north and south. Even though much of that sediment would be transported out the Golden Gate and only a portion of the remaining sediment would be available for bayland accretion, there could be enough sediment in a wetter future to allow Central Bay baylands to keep pace with the expected rate of SLR. It is less clear, however, if there would be enough sediment flux from the north and south under a drier future to support all baylands. The Richardson OLU has the potential for bayland sediment demand to be addressed in large part by local tributary sediment supply for both the wetter and drier future if that sediment can be directed towards and deposited on the baylands.

South Bay–Lower South Bay

Based on the findings from the future bayland sediment demand and sediment supply analyses, the South Bay–Lower South subregion could have a major deficit in the sediment supply needed to maintain the elevation of existing and planned restored baylands to 2100. Our analyses showed bayland sediment demand between 2050 and 2100 being at least five times higher than supply for the wetter future and at least ten times higher than supply for the drier future. The modest increase in the Alameda Creek sediment supply for the wetter future, and considerable decrease in supply for the drier future, help account for this difference in sediment deficit.

As with the other subregions, there are baylands with a high potential for long-term resilience (with respect to elevation keeping pace with SLR) given the right management approaches. The Colma–San Bruno OLU in South Bay has low bayland sediment demand to local tributary sediment supply ratios for the wetter and drier futures and therefore has a high resilience potential if that sediment deposits on the baylands. The Mowry and San Francisquito OLUs in Lower South Bay have low local bayland sediment demand to local tributary sediment supply ratios for the wetter future and moderately low ratios for the drier future, suggesting these baylands could also have a high resilience potential with local tributary and Bay sediment. The Stevens OLU in Lower South Bay could also have a high resilience potential with a combination of local tributary and Bay sediment if the polders are filled before restoration. However, it is important to note that if the future average annual net sediment flux direction is from South Bay to Lower South Bay, as it has been in the recent past (Livsey et al. 2020), Lower South Bay baylands could receive the sediment needed to maintain their elevation under a rising sea level. However, this would result in less sediment available for deposition on South Bay baylands.

Figure 3.5. (facing page) Map indicating baylands with the highest potential for resilience with respect to vertical accretion for a drier or wetter future based on the findings from this study and associated assumptions used in the analyses (Data sources: Schile 2012, SFEI and SPUR 2019).

Touring North Richmond during a king tide (Photo by Shira Bezalel, SFEI)

4
Sediment Management and Monitoring Considerations

Introduction

This regional exploration of sediment supply and demand for the present and future intertidal zone of San Francisco Bay is designed to advance understanding of sediment management needs for the baylands. The previous chapter forecasts a severe lack of sediment under conventional approaches, namely reliance on the *de facto* Bay sediment supply and natural intertidal processes to restore and maintain marshes. Those conventional approaches are already starting to change, and this chapter can inform that process. Analysis of the sediment deficit points toward many potentially useful lines of research and suggests a new approach to sediment management will be needed to meet the sediment demands of a healthy baylands ecosystem.

This chapter assumes a starting point based on the community goal to restore up to 100,000 acres (~40,000 ha) of tidal marsh. Keeping in mind that the previous chapters analyzed sediment demand for 75,000 acres (~30,000 ha) of marsh and identified a shortfall, that goal may need to be revisited over time, as we shift sediment management strategies and monitor the outcomes. Chapter 5 acknowledges the challenges associated with prioritizing use of sediment when it is scarce. Baylands managers and decision makers can expect to work through issues related to goal setting and prioritization as SLR accelerates and the impacts add up. The imperative for more aligned policies and management is only increasing.

This chapter is organized as a road map, from guiding principles to monitoring, for developing a holistic approach to sediment management for baylands resilience. The road map proceeds across an arc of information and activities that can help guide the community interested in the persistence of tidal flats and marshes. The key steps on the road map are:

- Employ principles of managing sediment for baylands resilience—*Nine principles that integrate the latest science and promote forward-thinking management.*

- Understand management opportunities—*A content-rich section with quantification of potential sediment sources that are currently untapped.*

- Create a place-based sediment management strategy—*Step-by-step discussion of how to create a strategy for a given location.*

- Fill critical knowledge gaps—*A brief table of the most important scientific questions to answer to improve sediment management.*

- Monitor to track baylands resilience—*Discusses sediment management within regional frameworks for monitoring and regulation.*

Hayward Regional Shoreline (Ellen Plane, SFEI)

Employ principles of managing sediment for baylands resilience

Managing sediment at the scales needed to take advantage of natural processes in watersheds, the Bay, and the Delta will require reworking of the approach to management and coordination currently in place. For example, sediment management currently is rarely coordinated between watersheds and baylands, the Delta and the Bay, or a locality that begins to scour when accretion happens in another place. A set of principles that helps guide thinking and planning can aid the transition to a new approach.

The following set of nine guiding principles (A to I) are based on scientific investigations and community processes related to understanding the estuarine ecosystem and preparing for climate change impacts (e.g., Goals Project 2015, Milligan et al. 2016, SFEI and SPUR 2019). The list below distills the ideas from these other efforts to include only the aspects most relevant to sediment management.

(A) Complete tidal marsh systems are the goal.

The Baylands Goals 2015 Science Update recommended focusing on restoring and protecting complete tidal marsh ecosystems to improve their resilience and thereby provide high levels of their desired services. To realize the multiple benefits of these ecosystems, the full system needs to be restored. That means including "all the following components appropriate to the local setting: submerged aquatic vegetation (SAV) beds, oyster beds, algal beds, rocky habitats, beaches, mudflats, low marsh, marsh plain, high marsh, complex channel networks, and transition zones, including natural levees along channels, creeks, and waterways, and broad transitions to adjacent wetlands and uplands" (Goals Project 2015, p107).

Focusing on sediment, this principle means considering other components, beyond marshes, such as tidal flats and shallow subtidal as part of the Bay's sediment need. Tidal flats and shallows help maintain marshes by attenuating waves and acting as a temporary reservoir of sediment for waves and currents to resuspend redistribute onto marshes (Schuerch et al. 2019). Here we did not calculate a sediment demand from the shallows, because it was not clear how the subtidal bathymetry is expected to evolve as sea level rises.

(B) Geography matters; consider geography in planning.

The heterogeneity in shoreline types and conditions around San Francisco Bay and the Bay's landscape position as the lower half of a very large estuary means that there is no one-size-fits-all sediment management approach. The particular characteristics of each shoreline reach (e.g., tributary sediment supply, location along the salinity gradient, proximity to the Delta, extent of planned tidal marsh restoration, potential marsh migration space) are important to consider for baylands sediment management. Operational Landscape Units (SFEI and SPUR 2019) are a useful scale for considering physical processes and managing shore and bayland resilience. The findings from Chapter 3 of this report comparing sediment supply to demand at the OLU scale are important when considering the geography of place-based approaches. See section 4 of this chapter for details on creating a place-based strategy.

The Bay is only one half of the estuary and is strongly influenced by events in the other half — the Sacramento-San Joaquin Delta. The Delta provides 37% of the average annual sediment supply to the Bay (Schoellhamer et al. 2018). The extensive planned marsh restoration in the Delta and Suisun Bay would be an ongoing sediment sink, likely to reduce sediment supply to the baylands. Levee breaches in the Delta from floods and earthquakes could also cause sediment sinks that reduce Bay sediment supply.

(C) Expect change, plan for multiple future states.

The environmental conditions that prevailed over the past century are rapidly changing, so decisions must not be made based solely on recent experience. Rather, decision-makers need to use projections of future conditions to guide planning now: increasing the resiliency of functions and services and lessening impacts to people and wildlife.

A conservative approach to planning for change in sediment supply would be to plan for a moderately hot and dry future, because this has a lower sediment yield (e.g., the HadGEM2-CC 8.5 climate scenario used in this study). Strategies for sediment management could be constructed around this future climate, with phasing triggered by thresholds or timelines. As elements of the strategy are implemented, monitoring would help guide whether the climate trajectory is in keeping with the scenario, and adjustments could be made as necessary.

(D) Take advantage of natural processes, which means managing sediment and water together.

Natural processes can address the root causes of ecosystem degradation and bolster resilience by allowing ecosystems to evolve and continue to function in response to climate change (Beechie et al. 2010). In the case of tidal marshes, the natural process of sediment delivery by creeks and the tides can nourish marshes and tidal flats to help them accrete at pace with SLR. Terrestrial runoff and tidal flows are key drivers of sediment processes in tidal flat and marsh systems, so management of freshwater and tidal flows is inextricably linked to effective sediment management. For example, fine sediments delivered to marshes via creeks and the tide, in conjunction with marsh plants, create topographic heterogeneity and channel systems that support habitat complexity (Temmerman et al. 2005).

When restoration of natural processes isn't possible or practical, management actions could emulate natural processes. For example, a novel approach yet to be proven involves placing fine sediment along a sediment delivery pathway, in subtidal channels or in shallows adjacent to marshes, on a regular basis and in relatively small amounts, allowing waves and tides to resuspend and redistribute the sediment onto marshes. However, mechanically dredging and moving sediment to flats and marshes is expensive and logistically challenging.

(E) Don't take for granted current access to sediment, freshwater, and tidal flows.

Once considered waste products, discharge from wastewater treatment plants and dredged sediment are now seen as valued resources to be recycled and reused, thanks to the work of many agencies working together on this issue. As their value increases, environmental uses of freshwater and sediment may not be able to compete with other sectors. For example, much freshwater is already being recycled and reused within watersheds. While this is a good outcome in many ways, some freshwater needs to be reserved for use at the shore to create complex, resilient, and complete tidal marsh systems, for example, in some horizontal levees (an irrigated, vegetated low-slope ramp bayward of flood risk management levees and landward of a tidal marsh) (TBI 2013).

A similar situation exists for sediment and other materials that can be used to protect shorelines and to elevate buildings and roads in flood zones. Such materials will be in demand to augment tidal marsh and tidal flat elevations, and build transition zone habitats, including horizontal levees. As more and more SLR adaptation plans include elevating infrastructure, competition may arise between using these materials for private development (such as elevating buildings in a privately owned office park) and for projects for the common good (such as building a horizontal levee that provides flood protection, habitat, and water quality benefits). Appropriate policies can be developed now so that limited resources of sediment and freshwater can be used to maximum advantage along the shore.

Finally, there have been several proposals to block tidal flows with gates in order to protect the shoreline. It is important to point out that tidal barriers, though effective at reducing high water levels in the short term, impact the natural sediment processes that sustain marshes in the long run. In muted marshes, natural processes are diminished and cannot promote accretion at the highest elevations of the marsh plain, thus limiting the ability of the marsh to accrete at pace with SLR. In locations where tidal barriers have been employed, large ecosystem shifts have impacted estuarine functions (Smaal and Nienhuis 1992). The baylands management and regulatory community will likely need to continue to educate others seeking to protect the shoreline about the long-term consequences of tidal barriers, and work collaboratively to find more optimal solutions for flood risk management.

(F) Combine management approaches into strategies.

Estuaries are complex systems comprised of many different elements and processes, and are strongly influenced by external drivers from watersheds and the ocean. This means that managing estuarine sediment processes as a system, rather than as individual parts, should help maximize the value of investments in sediment management. Such a systems approach would be a change from the status quo: each entity managing sediment within their jurisdiction, often without coordination with other entities or sectors across the full system.

Below is a list of objectives important for a sediment management strategy focused on increasing accretion in marshes. Furthering these objectives would likely help marshes persist longer. Concurrent with this approach, a longer-term strategy may need to be developed that acknowledges that marshes may become narrower and squeezed up against the shoreline later in the century. Planning for migration space and managed retreat to create that space could be key elements of such a strategy.

The goal of increasing tidal marsh and tidal flat accretion can be achieved through coordinated, landscape-level management of several sediment objectives. **First, sediment conveyance by creeks can be improved with a variety of actions.** More naturalistic, pulsed flows from reservoirs can move sediment downstream without damaging habitat for anadromous fish (Yarnell et al. 2015) (Note that managing pulsed or functional flows is a complex science that requires appropriate reservoir management, floodplain space, and scientific understanding of aquatic wildlife populations and habitat). While constructing new dams or retrofitting existing ones, sediment bypass tunnels (Serrana et al. 2018) and other modifications could be incorporated, allowing the opportunity to spill water from the base of dams. Then, sediment will be carried downstream rather than fillig up the reservoir. Dam removal and dredging reservoirs to remove sediment and reuse it in the baylands are other ways to access upper watershed sediment resources.

Los Gatos Creek below Vasona Lake Dam (photograph by Don DeBold, courtesy of CC 2.0)

The second objective is to increase sediment delivery from creeks to baylands. Reconnecting creeks, especially those with large sediment loads, to baylands by realigning channels and removing levees is an important action that has the benefit of increasing reliance on local sediment sources (rather than relying on Bay and Delta sediment that is not under local control). Also, flood control channels and other infrastructure can be redesigned to improve flow conveyance and sediment delivery to downstream baylands. Finally, in areas where the tide is muted, restoring full tidal action will allow natural processes to move sediment up onto marshes, providing the best chance for accretion to keep up with SLR. Conversely, tidal barriers and other muting of the tide hinder the natural processes that allow tidal marshes to be adaptive and resilient to SLR. Such actions should be avoided, or valuable wetlands and their benefits will be lost over time.

While the previous two objectives have focused on unlocking inorganic sediment in watersheds and delivering them to baylands, **the third objective is to harness the power of plants to increase organic matter accumulation within tidal marshes.** This objective can be accomplished by using freshwater (runoff and wastewater discharge) in estuarine-terrestrial transition zones to support freshwater and brackish marsh development. These types of marshes accumulate peat faster than salt marshes, which can help marsh plain elevation keep up with rapid SLR. Creating freshwater and brackish transition zones along the backshore of marshes has the added benefit of replicating the complex, heterogeneous habitat mosaics that were found historically where creeks entered the bay. Some new low-slope levee designs (e.g., horizontal levee) incorporate treated wastewater and other freshwater inputs onto the levee slope to achieve multiple benefits. Reconnecting creeks and other freshwater into the backshore of marshes also helps support marsh types that can accumulate organic matter quickly.

The final objective is to directly place or otherwise reuse dredged sediment in the baylands. In San Francisco Bay, beneficial reuse of dredged sediment has largely been used to bring subsided areas up to appropriate elevations prior to restoring tidal action. In addition to that practice, other placement approaches could increase the elevation of existing marshes, either through thin layer placement of sediment onto the marsh plain and in the estuarine-terrestrial transition zone (as was done at Seal Beach National Wildlife Refuge in southern California in 2016) or by placing sediment in areas where tides, fluvial flows, or currents would be likely to distribute them in a more natural topographic pattern on marshes and tidal flats. It may also be advantageous to consider sediments of all grain sizes—not just fine-grained sediments from navigational dredging—for beneficial reuse within the baylands. For example, sand and shell fragments could be used to augment tidal flats comprised primarily of these materials. It is important to consider detrimental effects from sand and shell mining, and the intent is to consider what sand and shell should be used for, not to increase the amount mined from the Bay. In some cases, upland soils or dirt might be useful for raising elevations as well, specifically as non-cover material or at the landward edges of marshes grading into transition zones. All the novel approaches mentioned above require study to ensure they are environmentally safe and do not cause unintended consequences.

(G) Collaboratively implement strategies at the landscape scale.

New relationships, cultural change, and unprecedented coordination will be necessary to implement the principles above and manage sediment holistically. The many overlapping jurisdictions and lack of alignment across policies and agencies means that system-scale management of sediment cannot happen until the entities that manage each element of the system (reservoirs, floodplains, creeks, flood control channels, wastewater treatment plants, infrastructure along the shore, baylands, and the Bay) coordinate with each other to optimize functioning and maximize benefits. This is a very tall order and will be a challenge to achieve before climate change accelerates and begins to foreclose options that are currently possible.

Organizations that are committed to baylands restoration and resilience need to reach beyond the sediment supply opportunities solely from navigational dredging. Dredged sediment is important and useful, but the volumes currently reused for restoration are only part of the answer to the problem. Even with the highly coordinated policy on dredged sediment, large quantities on an annual basis are not being reused due to federal policy limitations. Significant effort needs to move toward accessing watershed sediment supplies. Flood control agencies, water utilities, counties, cities, wetlands restoration organizations, other land managers, and regulatory agencies need to find creative solutions for holistic, system-scale sediment and freshwater management from reservoirs to the Bay.

Appropriate scales for coordinating management of sediment and freshwater are watersheds (for most freshwater sources and the sediments they carry) and baylands OLUs (for tides and many localized freshwater sources, like wastewater discharge). Upland dirt and dredged sediment may be transferred across the boundaries of watersheds or OLUs, although using material as locally as possible will reduce costs and pollution associated with transportation.

(H) Coordinate at the regional and estuarine scales.

In addition to the unprecedented coordination that is needed within watersheds and shoreline OLUs to improve sediment management, coordination across the region and the full estuary must be continued and augmented. Decisions made at the local scale will affect others across the region. For example, restoration of a subsided area can cause a sediment sink, leaving less sediment available for a neighboring area and even eroding a tidal flat nearby. These interdependencies, and the limited amount of resources available (sediment, freshwater, money, time, the attention of elected officials) mean that the Bay and full estuary community need to align with each other in how allocating these resources.

The entities engaged in bayland restoration and management have a history of coordination in the Bay, but coordination across the full estuary is less robust. The Baylands Ecosystem Habitat Goals processes (Goals Project 1999, 2015) established goals and recommendations for restoring baylands. As climate change drives environmental change, tidal marsh restoration continues, and more is learned about how decisions in one area affect outcomes in another (Wang et al. 2018), this regional coordination needs to continue, and goals may need to be recalibrated.

For the same reason, it is necessary to strengthen coordination across the Bay and Delta. Delta sediment supply to the Bay is influenced by management decisions, like extensive wetlands restoration in the Delta , whether levees are allowed to fail and create sediment sinks, and if a water conveyance system reduces flows to the Bay during large precipitation events. As the downstream recipient of decisions made in the Delta, the Bay community would be wise to seek stronger ties to Delta decision makers, to both influence and prepare for change in the Delta. Beyond these Bay-focused concerns, there are various important ecosystem functions and processes for which better coordination across the full estuary would be beneficial to all parties (including support for migratory and wide-ranging wildlife and migration of habitats with climate change).

(I) Learn rapidly with a robust monitoring and adaptive management program.

The importance of a robust sediment monitoring and adaptive management program is paramount. Environmental conditions are changing and will begin to change more rapidly, based on climate change model predictions. New approaches for living shorelines and other climate adaptation solutions are advancing rapidly. Ideally, this report could be one stone in the foundation of a sediment monitoring and management program with frequent analytical updates (perhaps every 5 years) and ongoing coordination across the entire estuary. Such a program would be one way to combat the inherent uncertainty in preparing for climate change impacts on the estuary. See section 6 in this chapter for a discussion of the relevant regional monitoring programs.

Coyote Creek (Photo by Don DeBold, courtesy CC 2.0)

Palo Alto Baylands at low tide (Photo by Don DeBold, courtesy CC 2.0)

Understand management opportunities

Without new management approaches, natural sediment supply from the Delta and Bay tributaries could fall short of what is needed for existing and planned baylands habitat to keep pace as sea level rises. There are currently many in-Bay and watershed sediment sources around the region that are locked in place or exported out of the system as waste material. However, many of these sources could be considered reserves for baylands restoration and long-term support if management approaches shift. Similarly, new approaches to treated wastewater discharge at the shoreline could augment local organic matter accumulation, helping increase bayland elevation. In addition, there are a range of watershed and bayland management actions that need to be implemented to support and augment natural sedimentation processes and increase the amount of sediment delivered to and deposited on baylands. This section provides a quantification by source of the amount of in-Bay and watershed sediment that could be used to support baylands, and an overview of key watershed and bayland management measures (or actions) that could increase the amount of sediment getting out of watersheds and onto baylands.

Resources that could be utilized to support bayland resilience

Within the region, waste products are being viewed more and more as valued resources for bayland restoration and shoreline adaptation, causing a shift in thinking around how to manage these resources to address bayland sediment deficits and accelerate bayland peat production. While some materials that would likely be disposed of—such as dredged sediment and excavated soils—are already being used in baylands restoration, additional opportunities exist to support bayland resilience. Resources that could be used to increase organic and inorganic accretion include (1) **tidal and fluvial sediment** dredged or mined from the Bay, stored in and removed from flood control channels, and trapped behind dams; (2) **upland sediment** in the form of excavated soils that currently end up in landfills, biosolids generated from wastewater treatment plants, and construction and demolition waste (e.g., brick, concrete, masonry, gravel, and stone) that could be used as a polder fill alternative in adaptation planning to preserve mineral sediment for more direct biotic uses; and (3) **treated wastewater** as irrigation in horizontal levees to support organic matter accumulation in marshes. Although biosolids, construction/demolition waste, and treated wastewater via horizontal levees are not permissible at present, these resource streams could be viable in the future as research evolves and sediment deficits increase. For each opportunity, we quantify the known relative magnitudes when available and provide other salient details. Refer to Appendix C for supplemental information used to calculate sediment mass for each source.

If the average annual production of sediment excavated from flood control channels, trapped behind dams, dredged from the Bay, and discarded to landfills (i.e., upland soils and biosolids) continues as it has in the past, there would be a total of ~600 Mt of sediment generated between 2010 and 2100 that theoretically could be available for reuse in the baylands (Figure 4.1). When compared with projected bayland demands, this is almost twice the amount of sediment that is needed for the existing baylands to keep pace with 6.9 ft (2.1 m) of SLR by 2100, and surpasses the amount of sediment needed for all additional restoration to be successful and keep pace with SLR through the end of the century. In comparison to projected sediment supply, 600 Mt of additional sediment is over two times more sediment than is projected to be supplied from the Delta and Bay tributaries under a wetter future and almost four times the amount of sediment projected to be supplied under a drier future (see Chapter 2 for more details on demand and supply projections).

About 50 Mt of the 600 Mt sediment projection is in reservoirs behind dams, representing a stockpile of sediment that could be used to fill polders or create horizontal levees or ecotone levees (i.e., non-irrigated horizontal levees). The remaining 550 Mt of sediment would be in the form of ~6 Mt of sediment generated annually on average across all sources described above (Figure 4.1). This sediment could be used to augment

Estimated future baylands sediment demand (2010-2100, Mt)*

		Legend
Existing baylands	364	▮ Tidal marsh
Existing baylands plus restoration	548	▮ Tidal flat
		▨ Polder fill

Estimated future Bay sediment supply from local tributaries and the Delta (2010-2100, Mt)*

Drier future:	159
Wetter future:	278

Tidal and fluvial sediment	**Historical (brown) & projected (black) cumulative mass of sediment (2010-2100) (Mt)***	**Current destination**
Navigation dredging	323 (~100% of Bay dredging)	Beneficial/upland reuse, waste product
Removed from flood control channels	11 proportion likely unaccounted (black bar based on ~12% of Bay Area channels; gray bar estimates remaining 88%)	Beneficial/upland reuse, waste product
Stored in flood control channels	-- no regional data	Stationary
Trapped behind dams	117 49 68 (~97% of dams in Bay watersheds)	Stationary

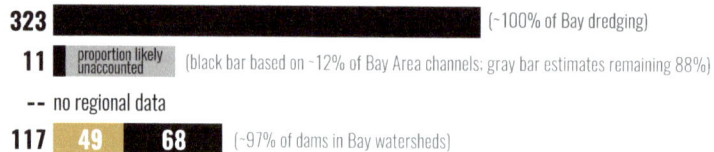

Upland Sediment	**Projected cumulative mass of upland materials (2010-2100) (Mt)**	
Excavated soils	151 (Landfilled portion only, unknown % reporting)	Waste product
Construction and demolition wastes	-- no regional data	Waste product
Treated biosolids	3 (~100% reporting, landfilled and other)	Waste product, agriculture

*Based on findings described in Chapter 2
**See Appendix A for densities used to convert volumes to mass for each category

the supply from Bay tributaries and the Delta for bayland habitats to keep pace with 1.9 ft (0.6 m) of SLR in the near term (2010–2050) and an additional 5.0 ft (1.5 m) of SLR in the long term (2050–2100). It could also be used to incrementally fill the polders slated for restoration to tidal marsh elevations or build horizontal levees or ecotone levees.

It is important to recognize the uncertainty in the projected magnitudes of the sediment sources described above. For instance, 600 Mt of sediment is likely the lower end of what could be generated through the end of the century since no data were obtained to estimate the amount of sediment trapped behind the 52% of unaccounted-for dams, excavated from the 88% of unaccounted-for flood control channels, and stored in all of the flood control channels that drain to the Bay. Furthermore, the magnitude of upland soils quantified (~151 Mt) reflects only the portion that went to regional landfills, so the true magnitude of upland soils stored in stockpiles or otherwise disposed of is unknown. Conversely, the numbers reported could be an overestimate when accounting for usability with respect to contaminant loads, grain-size requirements, or other factors that may limit what can be applied in the baylands. Uncertainty also exists around the future average annual trends for each resource. Competition for sediment could increase as regional adaptation efforts progress and lead to fewer opportunities to reclaim sediment disposed of as a waste product. In addition, several factors should be noted such as the bulk

Figure 4.1. Comparison of regional sediment reuse opportunities with projected bayland sediment demand and tributary sediment supply from 2010 to 2100. The 49 Mt of sediment trapped behind dams is the total amount accumulated as of 2008, representing a stockpile of sediment that could be used now to fill polders. The black bars denote cumulative mass of sediment that would be generated between 2010 and 2100 if average annual rates from the recent past continue. The black bars also denote sources that could be used to augment sediment supply from Bay tributaries and the Delta for bayland habitats to keep pace with 1.9 ft (0.6 m) of SLR in the near term (2010–2050) and an additional 5.0 ft (1.5 m) of SLR in the long term (2050–2100).

Projected trend	Avg. annual (Mt)	Assumed bulk densities	Time period	Data source	Potential use	Notes
↓	3.6	1.55 t/m³	2008-2017	LTMS 2019	Polder fill, placement	
↑		1.66 & 0.80 t/m³	~1973-2013	SFEI 2017a	Polder fill, placement	Based on data from ~12% channels
-	--	--	--	--	Polder fill, placement	
↓	0.7	1.19 t/m³	~1944-2008	Minear and Kondolf (2009)	Polder fill, placement	Projections do not account for dam storage capacity
↑	1.7	n/a	2006-2017	R. Egli, pers. comm.	Polder fill	
↑	--	--	--	--	Polder fill	
↑	<0.1	n/a	2009-2018	G. Kester, pers. comm.	Placement	Based on assumption of 20% solids

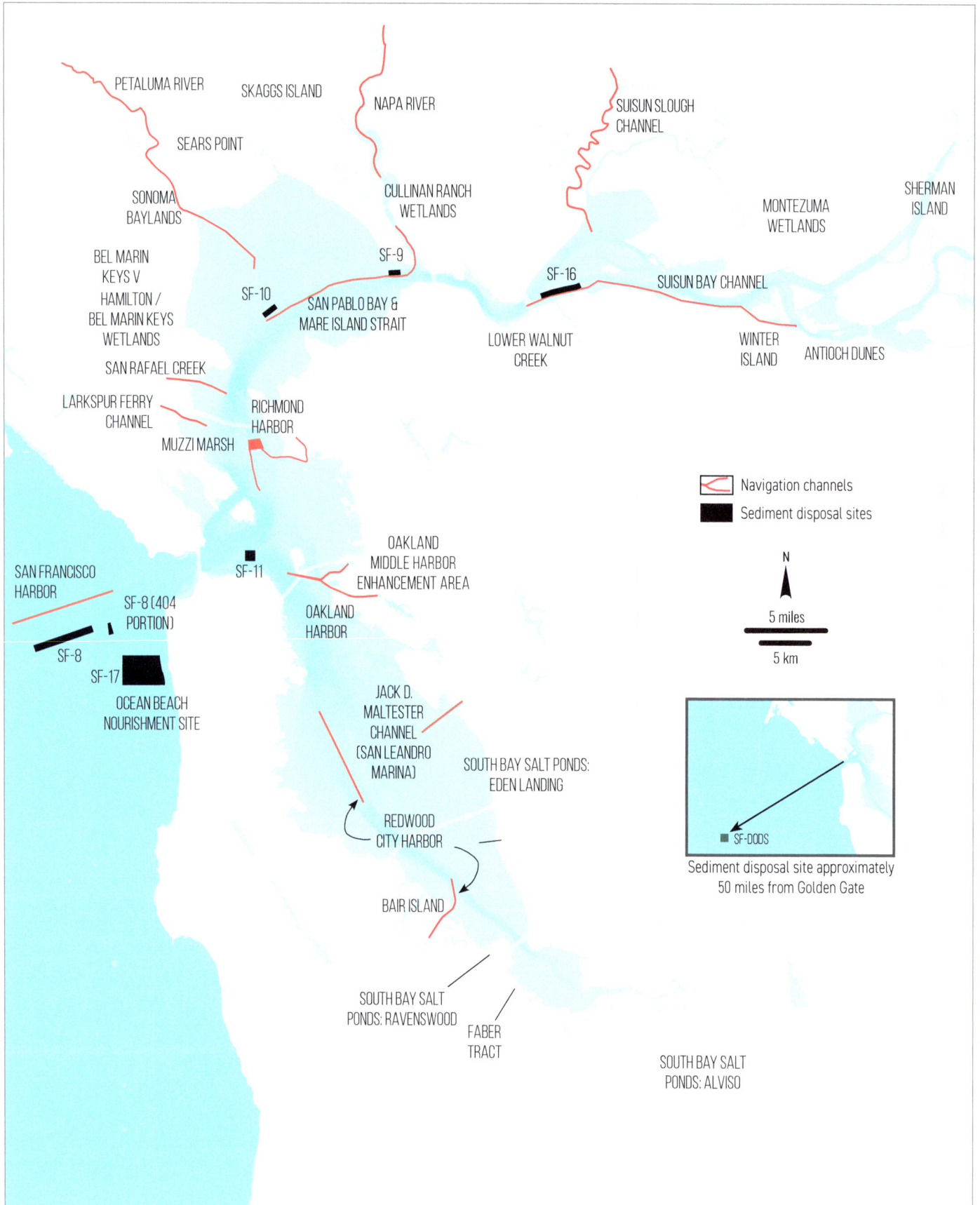

PETALUMA RIVER

SKAGGS ISLAND

NAPA RIVER

SUISUN SLOUGH
CHANNEL

SEARS POINT

SHERMAN
ISLAND

SONOMA
BAYLANDS

CULLINAN RANCH
WETLANDS

MONTEZUMA
WETLANDS

BEL MARIN
KEYS V
HAMILTON /
BEL MARIN KEYS
WETLANDS

SF-9

SF-10

SF-16

SAN PABLO BAY &
MARE ISLAND STRAIT

SUISUN BAY CHANNEL

LOWER WALNUT
CREEK

WINTER
ISLAND

ANTIOCH DUNES

SAN RAFAEL CREEK

LARKSPUR FERRY
CHANNEL

RICHMOND
HARBOR

MUZZI MARSH

Navigation channels

Sediment disposal sites

OAKLAND
MIDDLE HARBOR
ENHANCEMENT AREA

N

SF-11

SAN FRANCISCO
HARBOR

OAKLAND
HARBOR

5 miles

SF-8 (404
PORTION)

5 km

SF-8

SF-17

OCEAN BEACH
NOURISHMENT SITE

JACK D.
MALTESTER
CHANNEL
(SAN LEANDRO
MARINA)

SOUTH BAY SALT PONDS:
EDEN LANDING

SF-DODS

Sediment disposal site approximately
50 miles from Golden Gate

REDWOOD
CITY HARBOR

BAIR ISLAND

SOUTH BAY SALT
PONDS: RAVENSWOOD

FABER
TRACT

SOUTH BAY SALT
PONDS: ALVISO

densities used to convert volumes to mass and the time periods used to derive average annual mass estimates for each resource. The economics, logistics, and regulatory issues related to each sediment source also vary considerably and present different challenges.

Tidal and fluvial sediment

- **Sediment dredged from the Bay for navigation:** Since 1990, dredging to maintain navigational channels, refinery wharfs, harbors, small marinas, and other maritime features in the Bay has generated an average of over 3 Mcy (2.3 Mm^3) of sediment each year (Moffatt & Nichol 1997, Foley et al. 2019, LTMS 2019), equating to around 3.6 Mt/yr assuming a bulk density of 1.55 t/m^3 (2,700 lb/cy) for dredged sediment (SSFBS 2015). Between 2008 to 2017, approximately 41% of dredged sediment (a total of ~12.4 Mcy (~9.5 Mm^3) or roughly 14.7 Mt) from the Bay was reused for baylands restoration or other upland activities such as stabilizing levees, construction fill, or capping and lining sanitary landfills (SFBRWQCB 2000, LTMS 2018, Foley et al. 2019). The remaining 59% was disposed of offshore or in the Bay, with about 39% (~6.0 Mcy (~4.6 Mm^3) or ~7.1 Mt) dumped at in-Bay disposal sites and about 20% (~11.7 Mcy (~8.9 Mm^3) or ~13.9 Mt) dumped outside of the Golden Gate into the Pacific Ocean (Figure 4.2, LTMS 2018). If the average annual dredging volume is similar in the future, approximately 1.8 Mcy (~1.4 Mm^3) or 2.1 Mt of additional sediment could be diverted from in-Bay and ocean dumping sites for reuse in baylands restoration each year (assuming a bulk density of 1.6 t of sediment per m^3 (2,700 lb/cy) of dredged sediment (SSFBS 2015)).

  Barriers to beneficial reuse are the availability of offloading equipment at the restoration sites, funding for operational expenses to place the dredged sediment,and a lack of restoration sites permitted to receive sediment (Foley et al. 2019). The timing between dredging and placement also poses challenges, often necessitating stockpiling sediment which requires additional transport and storage space. Contamination is also a limiting factor: PCBs, pesticides, mercury, and other heavy metals are bound to sediments and mobilized during dredging activities (Foley et al. 2019). Different considerations exist depending on the elevation of polder fill with respect to the biotic zone. The maximum depth of biological activities in marshes is approximately 3 ft (0.91 m), so dredged sediment placed within this zone—referred to as surface material—must be below ambient contaminant levels (per SFBRWQCB 1998). Dredged sediment placed below the biotic zone, known as foundation material, is relatively isolated from biological receptors and thus higher concentrations of contaminants are allowed within the "non-cover" zone.

Figure 4.2. (left) Locations of dredging to maintain and expand navigation channels in the Bay (red lines). Sediment is not slated for beneficial reuse is disposed of in the Bay or outside the Golden Gate (black) (Courtesy of Foley et al. 2019; data adapted from LTMS 2018).

- **Sediment in flood control channels:** Over the past 200 years, the channels that drain to the Bay have been extensively altered, causing sediment to become trapped in flood control channels and necessitating mechanical removal by flood control agencies and other entities. In the 40 year period between 1973 and 2013, approximately 5.8 Mcy (~4.4 Mm^3) of sediment were removed from around head of tide in 33 of the largest flood control channels (Figure 4.3) (SFEI-ASC 2017a). This equates to approximately 4.9 Mt of sediment (assuming a bulk density of 0.8 t/m^3 (~1,300 lb/cy) for tidal sediments (Porterfield et al. 1961) and a bulk density of 1.7 t/m^3 (~2,800 lb/cy) for fluvial sediments (NHC 2004)). Around two-thirds of the 4.9 Mt of sediment came from tidal reaches downstream of head of tide and about one-third came from fluvial reaches just upstream of head of tide (SFEI-ASC 2017a). Most of the sediment came from dredging activities that occurred on average every five years. Sediment removed from the fluvial portions of channels near head of tide tended to be somewhat coarser than the sediment removed from the tidal reaches.

Additional sediment removal occurs within flood control channels further upstream, but regional estimates are not available. The magnitude of sediment collectively removed from flood control channels across all nine Bay Area counties is likely very large and could expand the amount of sediment available for reuse in a significant way. Within Santa Clara County, for example, the Santa Clara Valley Water District removed around 1.4 Mcy (1.0 Mm^3) of sediment between 1978 and 2004 (SCVWD 2005). This equates to around 1.7 Mt of sediment, when converted using a bulk density of ~1.7 t/m^3 (~2,800 lb/cy) (NHC 2004) (Figure 4.3).

There is also sediment deposited within flood control channels that is not removed and remains stored within the channel. Tracking the amount of in-channel sediment storage entails conducting regular longitudinal and cross-sectional surveys to quantify long-term trends in channel elevation change over time or through comparison of recent surveys to as-built channel dimensions (SFEI-ASC 2017a). Currently, there are very few flood control channels being monitored in this way and no database exists to compile this type of information at the regional scale. The magnitude of sediment trapped in flood control channels varies greatly between channels, ranging from less than 0.001 Mt in Alhambra Creek to around 2.4 Mt in Alameda Creek Flood Control Channel (assuming a bulk density of about 1.7 t/m^3 (~2,800 lb/cy) to convert 1,145 cy (875 m^3) and 1.9 Mcy (1.45 Mm^3) of excavated volumes to mass respectively (NHC 2004)) (SFEI-ASC 2017a).

Some of the barriers to reusing sediment removed from flood control channels include high costs for transport and placement on restoration sites (refer to the dredging section for more information), high levels of contaminants, and a lack of regional coordination to match agencies disposing of sediment with restoration agencies in need of sediment. Although some of the sediment removed from flood control channels around head of tide goes to bayland restoration projects, over 60% ended up in landfills or disposed of as a waste product over the past several decades (SFEI-ASC 2017a), likely due to a combination of the barriers listed above. Sediment removed further upstream also likely ends up as a waste product or, in some cases, is stored in a stockpile and available for reuse (e.g., the Wildcat Creek Sediment Stockpile in Contra Costa County).

Boundary of the portion of Santa Clara
County that drains to the Bay

Flood control channel

Sediment removed (cubic yard/linear foot (1978-2004))

<5

5 - 10

10 - 25

25 - 50

50 - 80

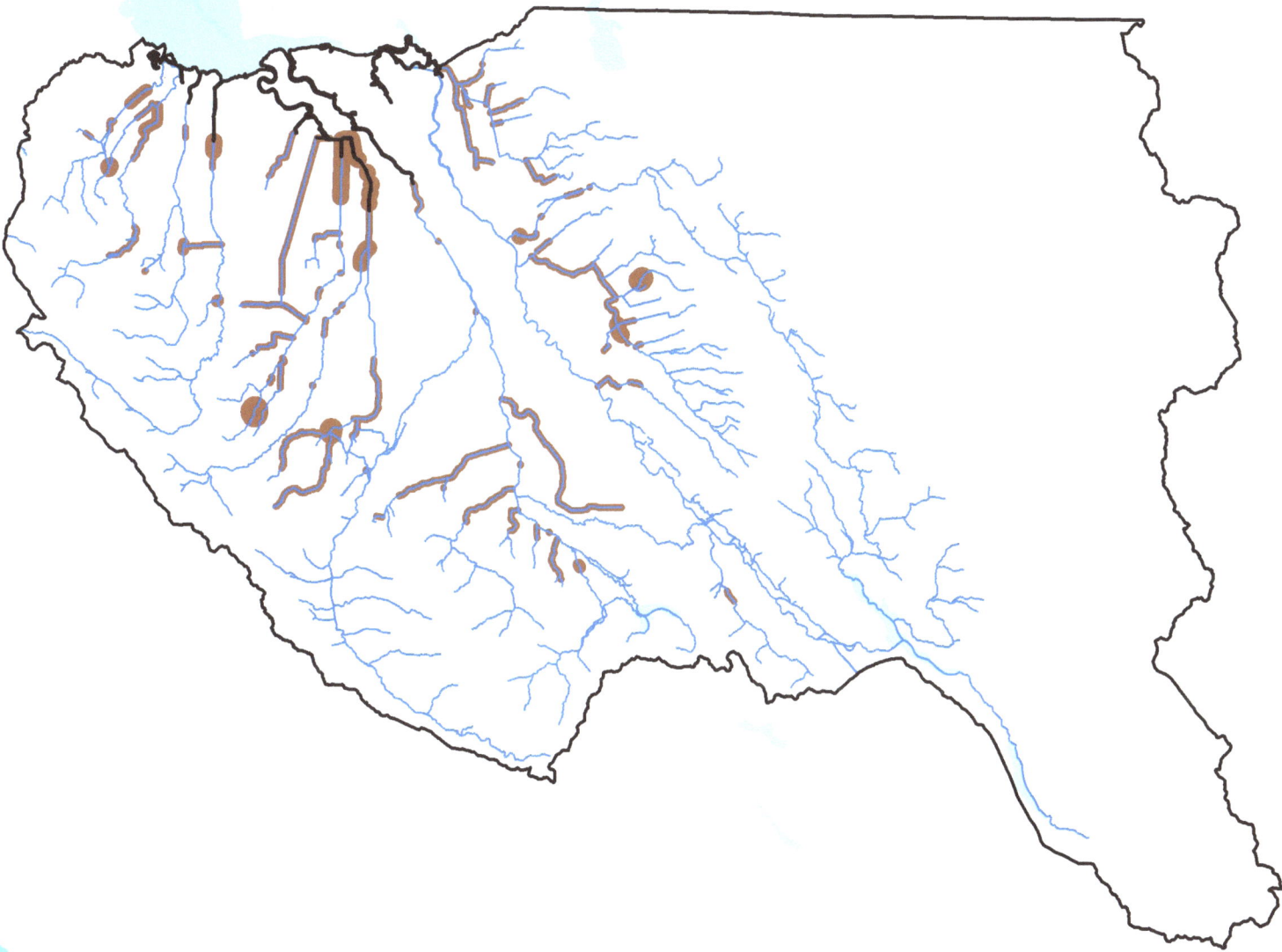

Figure 4.3. Locations and magnitude of sediment
removed from flood control channels in Santa Clara
County that drain to SF Bay between 1978 and 2004
(SCVWD 2005).

N

5 miles

Stevens Creek Reservoir (Photo by Don DeBold, courtesy CC 2.0)

- **Sediment trapped behind dams:** Dams dramatically disrupt the flow of water, sediment, and organic matter downstream (Roni and Beechie 2013). Over time, reservoirs behind dams fill with sediment which is costly to remove and can lead to numerous problems such as frequent spilling and backwater flooding upstream (Morris and Fan 1998). Since the 1850s, over 150 dams have been constructed in the watersheds that drain to the Bay for water supply and flood management, with the majority of dams built in the 1950s (DSOD 2020). As of 2008, approximately 41 Mm^3 (~54 Mcy) of sediment is estimated to be trapped behind 146 of the dams in Bay watersheds (Figure 4.4) (Minear and Kondolf 2009, Milligan et al. 2016). Using a bulk density of 1.19 t/m^3 (2,000 lb/cy) (NHC 2004), this equates to approximately 49 Mt of sediment (Figure 4.4) which could fill roughly 44% of the polders within in-progress and planned tidal marsh restoration sites. This estimate does not include sediment accumulation over the past 12 years, so the actual amount of sediment behind all Bay Area dams could be slightly higher. See The Hidden Sediment Reserve online data visualization tool for estimated costs (both financial and environmental) associated with getting reservoir sediment down to the baylands.

Like the other tidal and fluvial sediment sources discussed above, reusing sediment trapped behind dams has its challenges. For example, the costs of excavating and transporting accumulated sediments can be prohibitively costly and, in some instances, have been the largest component of nationwide dam decommissioning costs (Minear and Kondolf 2009). In addition, trucking excavated reservoir sediment to downstream baylands can result in considerable greenhouse gas (GHG) emissions. Mobilizing trapped sediment can also transport sediment-bound contaminants that can disrupt the biogeochemical processes within a stream and negatively impact wildlife (Hart et al. 2002). In addition, a sudden influx of sediment into a creek—such as immediately following a dam's removal—can increase the turbidity and sedimentation in the channels and impact fish and other aquatic species.

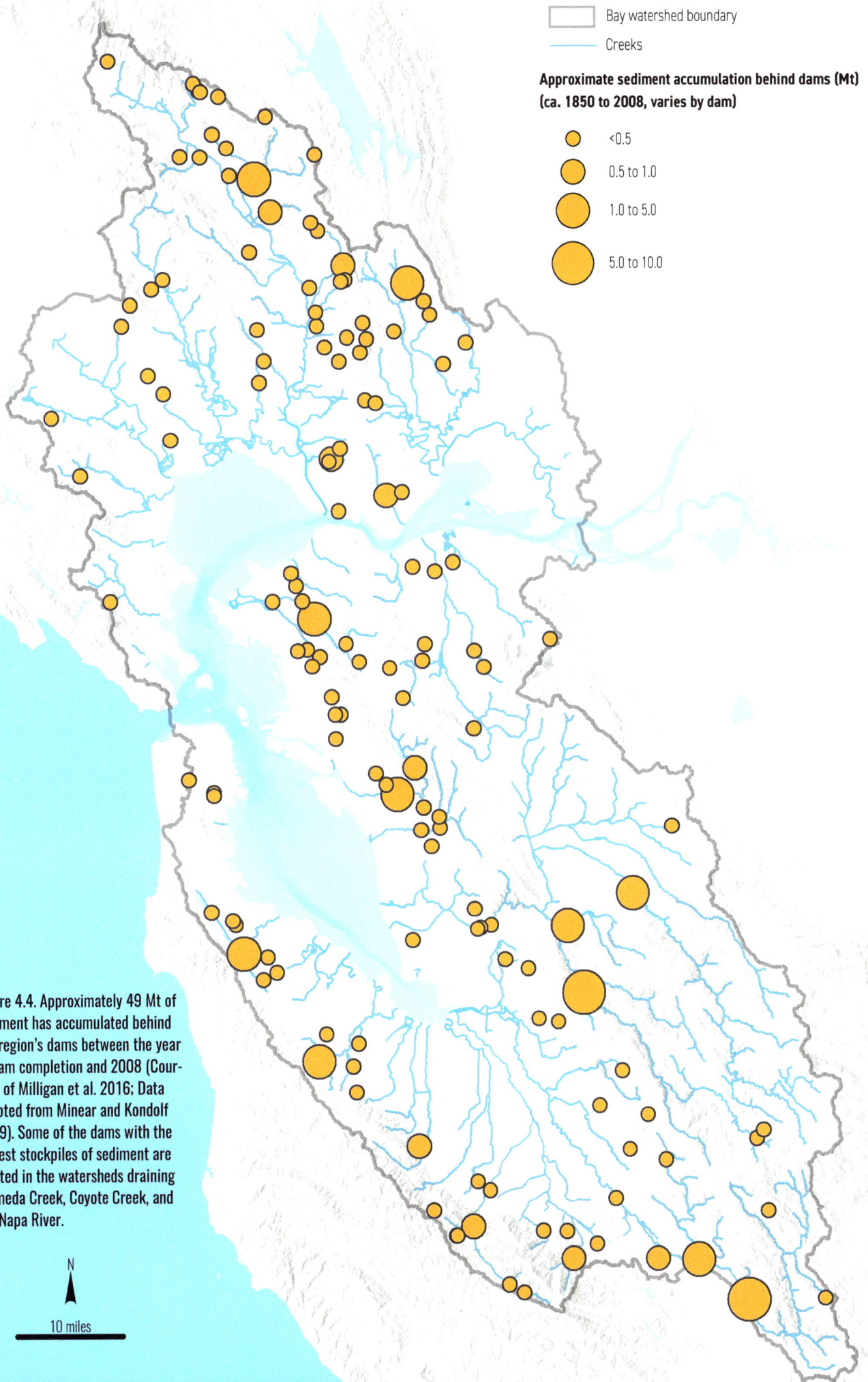

Figure 4.4. Approximately 49 Mt of sediment has accumulated behind the region's dams between the year of dam completion and 2008 (Courtesy of Milligan et al. 2016; Data adapted from Minear and Kondolf 2009). Some of the dams with the largest stockpiles of sediment are located in the watersheds draining Alameda Creek, Coyote Creek, and the Napa River.

Bay watershed boundary
Creeks

Approximate sediment accumulation behind dams (Mt)
(ca. 1850 to 2008, varies by dam)

<0.5
0.5 to 1.0
1.0 to 5.0
5.0 to 10.0

N

10 miles

Upland sediment

- **Excavated Soils:** The removal of soil (sediment, organic matter, and other materials) through excavation processes is a ubiquitous aspect of urban and suburban construction. Upland soil is removed for the building of foundations and the placement, maintenance, and expansion of numerous subsurface infrastructure systems (Bobylev 2015). In dense urban areas, the soil volumes yielded by these processes are generally transported off-site owing to spatial constraints limiting on-site reuse.

  Because of the considerable costs involved in stockpiling soil for reuse, it is often disposed of in large amounts in landfills, utilized as a "daily cover" application that is distributed over solid wastes to reduce windblown trash, scavenging, and odors (Price 2011) (Figure 4.5). As such, the "market" for managing excavated soils has been shaped by the regional solid waste management sector and contractors who physically transport soils from excavation sites to reuse or disposal sites.

  The mass-hauling of excavated soils for extensive landfilling is a major producer of GHG emissions, in addition to other negative environmental impacts (Magnusson 2019). Moreover, landfilled soil is not recoverable, becoming permanently "lost" from a resource reuse perspective when used as daily cover.

  Numerous data gaps hinder classification of various aspects of soils received by landfills, including geotechnical composition, legacy contaminant profiles and other aspects important to consider for reuse permitting and project-planning purposes. While some efforts to classify the impacts of soils reuse and recycling have been undertaken (Magnusson 2015), a general lack of strategic

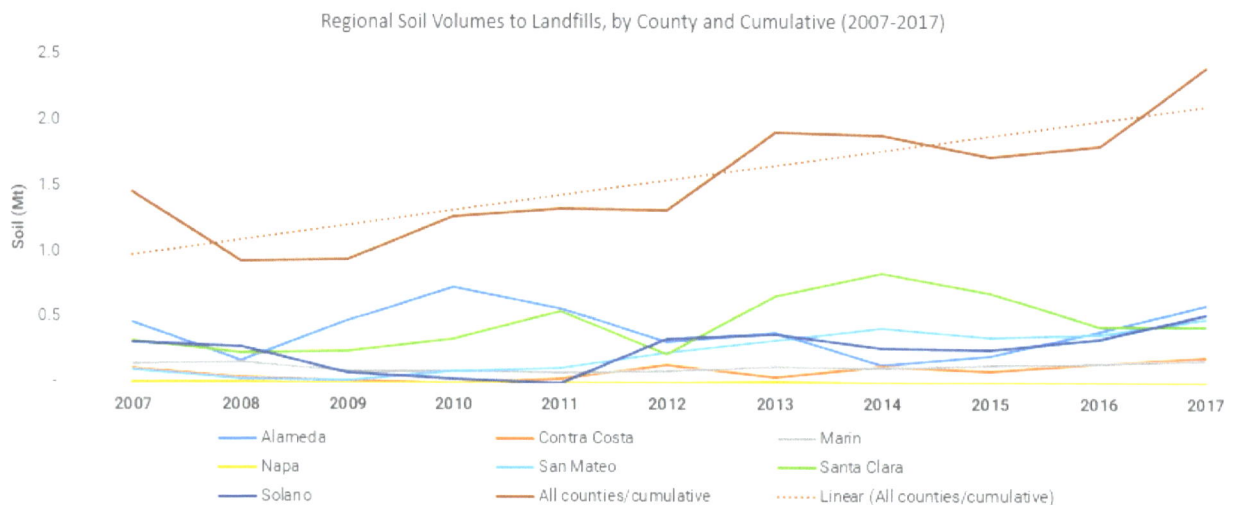

Regional Soil Volumes to Landfills, by County and Cumulative (2007-2017)

Figure 4.5. A survey of landfills in the region over a decade-long period between 2007 and 2017 reveals a total of nearly 17 Mt of soil deposited as daily cover, with the annual mass showing an increasing trend over time. The red line reflects a total mass of sediment disposed of at landfills by all nine Bay Area counties, and the dashed line indicates the regional trend for the time period analyzed. For comparison, the estimated sediment supply to the Bay from the Delta and the Bay tributaries between 2007 and 2016 was approximately 8 Mt (Schoellhamer et al. 2018). See Appendix C for details on the data source.

coordination for considering regional soils reuse persists. Moreover, the competitive marketplace shaping the reuse of excavated soils complicates their use in public sector projects and long-term planning processes. Nonetheless, because of the versatility and amount of this material being actively handled, greater consideration of management in the context of adaptation planning is appropriate, and will become increasingly important as sea level rises.

Construction and Demolition Wastes: Within a given region, enormous quantities of materials are produced as a function of the turnover in building and infrastructure stock (Hu et al. 2010). Construction and demolition wastes (CDW), representing perhaps one-quarter of California's total solid waste stream (R. Egli (CalRecycle), pers comm), are well-recognized as important opportunities for reuse by recycling and/or upcycling processes (Volk et al. 2019).

While excavated soils (discussed above) are generally classified as CDW (in some projects representing the overwhelming majority of material generated), certain other materials included in CDW flows possess high reuse potential. In particular, non-organically reactive components including brick, concrete, masonry, gravel, and stone could be reused to create subsurface mass in shoreline restoration and adaptation applications (e.g., base material in horizontal levees and restored polders). More research is needed to assess this potential.

- **Treated Biosolids:** Urban areas process and export solid human wastes on a continual basis and in volumes that correlate to the size of the population served. Between 2009 and 2018, wastewater treatment plants throughout the nine Bay Area counties generated approximately 1.5 Mt of biosolids collectively (water content is approximated at ~80% of total reported) (G. Kester, pers. comm.). Around 58% of this total ended up as waste products in landfills and the remaining portion—about 42%—went to agriculture and other uses (G. Kester, pers. comm.) (Figure 4.6).

Treated biosolids derived from human wastes are high in organic matter and nutrients, and are commonly reused as a land-application for agricultural purposes. After dewatering, biosolids are also disposed of in landfills as an "alternative daily cover" (ADC), and in some

REGIONAL BIOSOLIDS DISPOSAL: 2009-2018 (METRIC TONS)

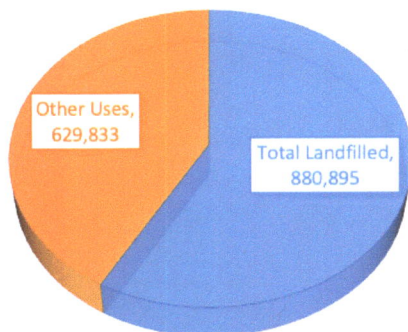

Other Uses, 629,833

Total Landfilled, 880,895

Figure 4.6. Between 2009 and 2018, Bay Area wastewater treatment plants generated approximately 1.5 Mt of biosolids collectively (water content is approximated at ~80% of total reported). Around 58% of this total ended up as waste products in landfills and the remaining portion—about 42%—went to agriculture and other uses (G. Kester, pers. comm.). See Appendix C for details on the data source.

cases are incinerated. As with the transport of large masses of soils, significant environmental impacts are incurred in the long-distance transport of these materials, in addition to the potential for methane off-gassing in landfills.

While well-recognized for upland soil conditioning, treated biosolids have also been shown to increase the vegetative growth of marsh macrophytes in the Bay when used as an amendment to dredged sediment (Foster-Martinez and Variano 2018). Atlantic coast salt marshes treated with wastewater-derived fertilizers have also been shown to exhibit increased primary and secondary productivity, as well as the long-term capacity to reduce nitrogen loading of adjacent water bodies (Brin et al. 2010).

Treated biosolids physically resemble soils in many respects, and might be considered as a simple source of mass for augmenting projects requiring bulk building materials. It is important to note that biosolids can contain concentrated levels of heavy metals, pharmaceuticals or other biologically active compounds that must be managed and accounted for in reuse applications, especially those where water quality is a concern (Lu et al. 2012).

Strategies in Applied Coastal Adaptation: Examples of Mass Material Reuse

The resources discussed in this section outline some of the particular materials within the suite of flows in and around the Bay Area. As the region reconsiders these resources and their potential roles in a future shoreline defined by rising sea level and a growing urban population, other cases of mass material reuse in coastal settings may provide useful comparisons.

Regions grappling with coastal zone hazards have utilized debris for coastal adaptation projects. Japan's catastrophic 2011 Tohoku earthquake and resultant tsunami generated over 20M tons of debris, which is being used and studied for reuse in engineered berms, embankments and levees in preparation of future tsunamis (Inui et al. 2012). Similarly, New York City's Office of Environmental Management led the collection of sand deposited by Superstorm Sandy for reuse in coastal berm construction.

Bair Island from the air. (Photo by Jitze Couperous, courtesy CC 2.0)

The application of enormous amounts of materials in shoreline projects is widely evident in San Francisco Bay, where baylands were filled to create vast swaths of flat, low ground. Indeed, debris from construction, demolition, and disaster areas—in addition to the wholesale landfilling of all manner of wastes at the shoreline—can play an important role in the region's restoration efforts. Within the modern regulatory frameworks, excavated soil is being sought and used in extensive shoreline restoration and development with SLR adaptation planning requirements in mind. The restoration of Inner Bair Island is a prime example of this, with over 1 Mcy (0.76 Mm^3) of clean upland dirt used to raise elevations before breaching to tidal action (BCDC 2017). The use of upland soils to help raise elevations at Inner Bair Island was pursued, in part, to decrease costs and quicken the pace of restoration compared to using just dredged sediment (SFBRWQCB 2008). Testing was required for all imported upland material to Inner Bair Island to ensure the materials met environmental quality standards (SFBRWQCB 2008). §

Treated wastewater for organic matter accumulation

Freshwater is a key input for tidal marshes and estuarine-terrestrial transition zones, delivering an influx of sediment and nutrients and supporting accelerated organic matter accumulation. Core samples from historic tidal marshes in the Bay indicate plant productivity can be higher in freshwater and brackish marshes compared to salt marshes (Schile et al. 2014). For example, marsh samples at China Camp, a salt marsh characterized by 10-30% water salinity, ranged from 150 to 1,750 g/m^2 (4.4 to 51.6 oz/yd^2) of organic matter, whereas samples from Browns Island, an oligohaline marsh with 0–5% water salinity, ranged from 160 to 3,200 g/m^2 (4.7 to 94.4 oz/yd^2) of organic matter (Schile et al. 2014). Many of the creeks that historically flowed to tidal marshes and their transition zones have been realigned or disconnected (SFEI-ASC 2017a), leaving large disparities in the amount of freshwater entering the baylands compared to historical conditions.

The concept of reclaiming freshwater from the urban landscape in the form of runoff and treated wastewater discharge to support marshes and their transition zones is in the early stages of implementation. For example, the Oro Loma Sanitary District in San Leandro, CA, is currently exploring the feasibility of using treated wastewater to irrigate a 0.7 ha (1.7 ac) horizontal levee. The concept of an irrigated horizontal levee was conceived in 2013 as a way to protect existing coastal infrastructure from storm surge while simultaneously filtering contaminants (Cecchetti et al. 2020). A recent study on Oro Loma's experimental system showed promising results in reducing a wide variety of wastewater-derived contaminants while rapidly establishing dense native vegetation gradients that attracted a range of wildlife (Cecchetti et al. 2020).

Treated wastewater is an abundant resource that could help build much-needed elevation capital in marshes and establish vegetated estuarine-terrestrial transition zones around the Bay's edge (Collins et al. 1999, Beyeler et al. 2015). Between 2017 and 2018, an average total volume of approximately 434 million gallons per day (MGD) (1.64 Mm3) of treated wastewater was discharged to the Bay from 45 wastewater treatment plants across 38 municipal dischargers (BACWA 2018, Davis et al. 2019, SFBRWQCB 2019). Nearly 50% of this daily average total was discharged to Central Bay, with the majority from three municipal dischargers: East Bay Dischargers Authority (59.7 MGD), East Bay Municipal Utility District (52.5 MGD), and the City and County of San Francisco Southeast (57.4 MGD) (Davis et al. 2019, SFBRWQCB 2019). About 24% was discharged to Lower South Bay at three outfall locations, one of which was the single largest wastewater treatment plant contributor in the Bay Area in 2017–2018 (San Jose-Santa Clara RWF, 87 MGD). Suisun Bay received about 13% of the average daily discharge, and the remaining was split between San Pablo Bay (~7%) and South Bay (~6%) (Davis et al. 2019).

Many challenges exist with the incorporation of treated wastewater discharge through the baylands and with the conversion of salt marsh to brackish or freshwater marsh (e.g., Collins et al. 1999). For example, pharmaceuticals and contaminants of emerging concern may not be filtered out by current wastewater treatment practices and and could pose ecological hazards through uptake by marsh plants and consumption by wildlife. Additionally, the volume of freshwater discharged from wastewater treatment plants could decrease with upgrades in water recycling technology (e.g., reverse osmosis; UCB et al. 2020).

Oro Loma (Photo by Ariel Rubissow Okamoto)

Watershed design and management opportunities
Flow management (landscape modifications)

Over the past 200 years, land clearing and urbanization in Bay Area watersheds have caused storm peak discharges to increase and storm hydrographs to become more flashy, meaning that there is a rapid increase in discharge over a short time period with a quickly developed peak discharge in relation to base flow (Ward 1978) (Figure 4.7). In many watersheds, these hydrograph changes have caused increased channel erosion, increased downstream sediment delivery, and increased in-channel fine sediment deposition associated with rapid discharge decrease following peak flow. At the mouths of some impacted watersheds, these changes have likely resulted in large amounts of watershed sediment being transported past tidal marshes and tidal flats where it was once deposited during small and modest storm events. By modifying storm hydrographs to more closely mimic historical conditions, it is possible to improve watershed channel stability and habitat conditions for sensitive local species while also maintaining a supply of fine sediment to downstream baylands and a flow regime that extends the time period during which storm flows are able to spread out over marshes and deposit fine sediment.

Modifying watershed hydrographs around the region to make them more similar to historical conditions should focus on floodplain restoration and low impact development (LID) implementation. Restoring broad floodplain areas above head of tide that would be inundated during frequently occurring floods (~2-year floods) and larger floods would allow flood flows to spread out and be temporarily stored, thereby decreasing the peak flow

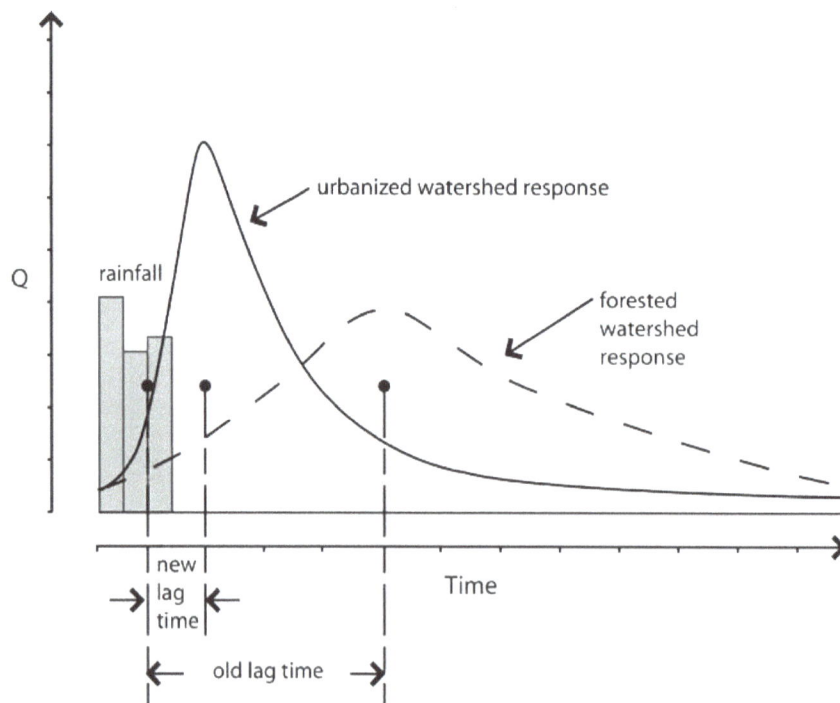

Figure 4.7 Conceptual watershed storm hydrographs showing the impact of urbanization on peak flow discharge, peak flow lag time, and overall hydrograph shape (adapted from Alberti 2008).

discharge, increasing peak flow lag time, and increasing the time the storm hydrograph tail is at a discharge level that allows sediment-laden storm flows to spread out over baylands. LID features (i.e., small-scale hydrologic control features that promote infiltrating, filtering, storing, evaporating, and detaining runoff close to its source) can be effective at capturing runoff and modifying peak discharge and lag time for more frequently occurring flood events (<2-year floods). These management actions should be explored in all highly urbanized watersheds with broad baylands at their mouths around the region, with a particular focus on watersheds with a high sediment supply relative to local bayland demand.

Flow management (dam release modifications)

In addition to widespread deforestation and urbanization, the building of dams around the region has modified flow and sediment dynamics. Dams built for water supply and flood management are effective at capturing and storing sediment-laden flood flows, thereby decreasing the magnitude and frequency of flood pulses and the sediment supply to downstream reaches (Grant et al. 2003, Magilligan and Nislow 2005) (Figure 4.8). Following channel incision and coarsening associated with changes to the flow and sediment supply, channel beds downstream of dams can become locked in place and embedded with fine sediment that is no longer flushed downstream. This can have negative impacts on a range of aquatic species including salmonids who need clean, mobile gravel for spawning. In the downstream baylands, dam-induced changes to flow and sediment dynamics can result in a decrease in long-term fine sediment supply. Modifying how flows are released from dams could therefore have benefits for both watershed and downstream bayland ecosystems.

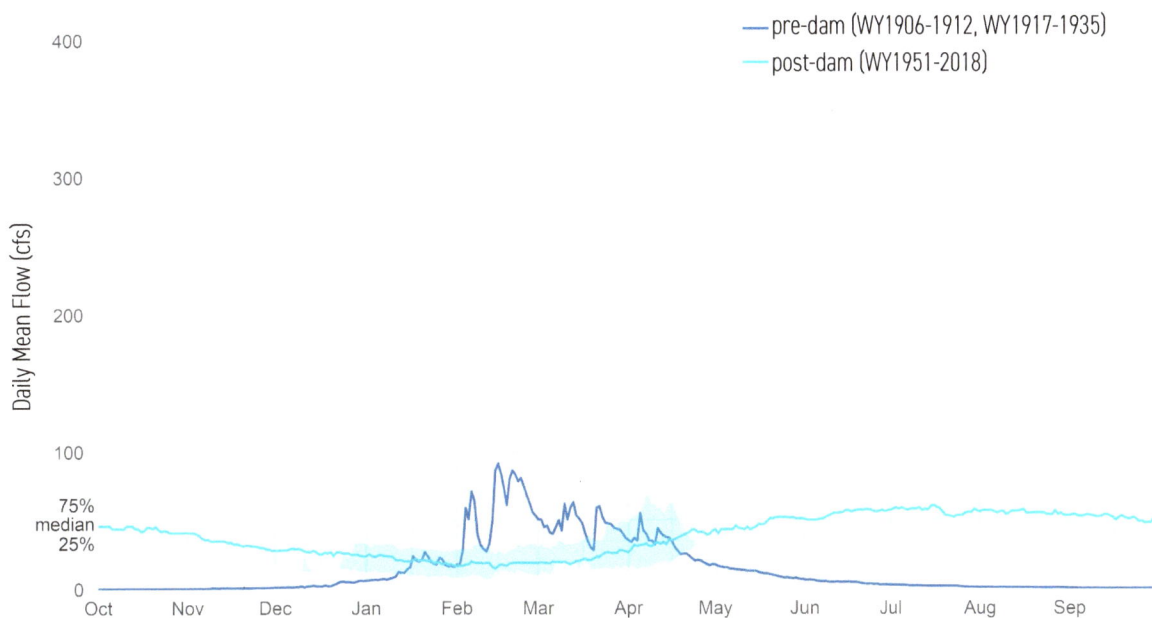

Figure 4.8. Daily mean flow downstream of Anderson Dam on Coyote Creek for pre-dam and post-dam conditions,

Where feasible, dam flow releases should include frequent large pulses during the winter and early spring that mimic historical conditions and transport fine bed sediment to downstream baylands. There are many examples of "flushing flows" being required for dams in the Sierra Nevada and throughout the western United States to mobilize downstream bed sediment and help mitigate dam impacts on downstream channel physical conditions and aquatic ecology (Kondolf et al. 1987, Gore 2018). Locally, implementing flushing flows has been recommended for dams that impound small and large reservoirs and irrigation ponds within the Napa River watershed to flush fine sediment from critical salmonid spawning habitat (SFEI 2012). Flushing flows should be designed for Bay Area dams as part of a larger "functional flow" approach that focuses on retaining specific process-based components of the wet-season and dry-season hydrograph and interannual flow variability rather than attempting to mimic the full natural flow regime (Yarnell et al. 2015). Often, existing dam infrastructure can prohibit implementing pulsed flows (e.g., the dam outlet pipe is too small to pass the desired peak discharge of the desired flushing flow) (Richter and Thomas 2007). In these instances, opportunities for infrastructure modification should be examined as part of seismic upgrades or other required reconstruction. Additionally, there are typically several management constraints that need to be assessed when considering the feasibility of flushing flows, including the potential to cause channel instability in sediment-starved reaches or flooding and infrastructure damage in highly developed reaches (Kondolf and Wilcock 1996).

Reservoir sediment management

As described earlier, major dams around the region collectively store more than 40 Mt of sediment. Efforts to move that sediment out of reservoirs to downstream channels and baylands where it can provide multiple benefits are now being explored around the region. For example, Alameda County Flood Control & Water Conservation District is currently exploring options for removing sediment from Don Castro Reservoir in the San Lorenzo Creek watershed. Created in 1964 with the building of Don Castro Dam, the reservoir is filled with mostly fine sediment, which has decreased its capacity to store floodwaters. The District needs to remove this sediment and is looking to do so in a cost-effective manner that includes releasing it to sediment-starved San Lorenzo Creek downstream and transporting it via slurry pipe to the bayland restoration site at Eden Landing (Public Sediment 2019). If implemented, this would be a first-of-its-kind project in the region that "unlocks" reservoir sediment in a way that maintains flood management priorities while supporting riverine and estuarine ecosystems.

Updated reservoir sediment management approaches that promote downstream sediment transport should be the long-term goal for all dams in the region as they are retrofitted. Proven techniques for passing sediment through or around reservoirs that should be considered include the following (from Kondolf et al. 2014):

- sediment bypassing—diverting sediment-laden flows around the reservoir and dam through a side channel or pipe (e.g., the Don Castro Reservoir described above)

Don Castro Reservior. (Photo courtesy of CC 2.0)

- <u>drawdown sluicing</u>—drawing down the reservoir and discharging sediment-laden high flows through the dam during periods of high reservoir inflow

- <u>drawdown flushing</u>—complete drawdown of the reservoir and scouring of the reservoir sediment deposit during high flows

- <u>pressure flushing</u>—reservoir levels are maintained and dam outlets are opened to scour reservoir sediment just upstream

- <u>turbidity current venting</u>—sediment-laden turbidity currents that enter the reservoir are allowed to pass through dam outlets

As with pulsed flows, opportunities for dam infrastructure modifications that are necessary for implementing new approaches for sediment routing should be examined as part of seismic upgrades or other required reconstruction. However, new approaches to reservoir sediment routing may not be feasible in many instances due to a variety of factors including cost and logistical constraints.

Creek realignment

Beginning in the 1850s, the lower, flatter portions of watersheds at the Bay edge were reclaimed for agriculture and development, which entailed realigning creeks with sharp angles to increase the extent of arable land and avoid discharging into salt ponds and building levees to help contain flood flows (e.g., Sunnyvale East and West Channels, San Francisquito Creek, Calabazas Creek). In the 20th century, the increased need to protec developed areas adjacent to these creeks from flooding led to channel extension, straightening, and widening, as well as the construction of engineered flood control levees. These channel modifications have caused the lower portions of many creeks to accumulate sediment upstream of sharp-angle channel realignments, requiring regular removal to maintain flood conveyance. There are dozens if not hundreds of examples of these problematic sharp-angle channel alignments spread throughout the region (Figure 4.9). Restoring historical alignments for those channels with excess sediment accumulation could increase sediment transport capacity, thereby improving flood conveyance and helping get more sediment to downstream baylands.

Opportunities should be identified around the region for channel realignments that provide multiple benefits, with a focus on the most problematic channels in terms of sediment accumulation and flood conveyance. Historical channel alignment (where known) should be used as a guide for redesign, acknowledging local constraints and landscape changes upstream and downstream that would necessitate a modified alignment. Many channels in need of realignment are located in highly urbanized areas, thereby limiting redesign options. However, channels in less-developed landscapes and in parks and open spaces within urbanized areas can provide opportunities for realignment that increases sediment transport capacity and improves flow conveyance in a manner that supports local aquatic habitat.

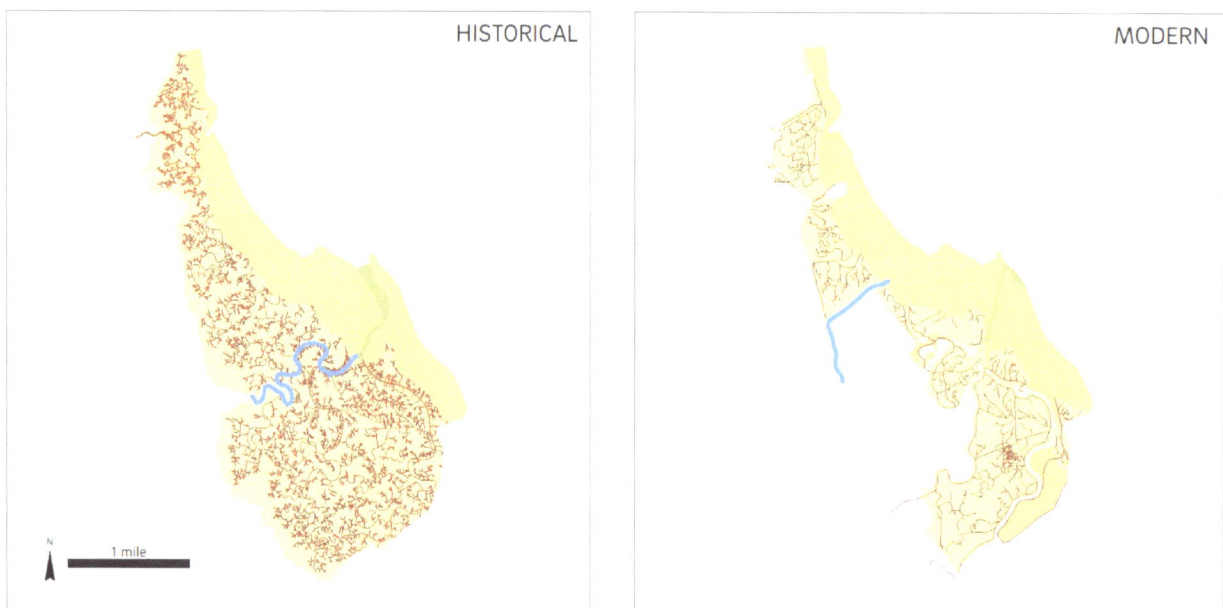

Figure 4.9. Historical channel realignment of lower San Francisquito Creek (source: SFEI-ASC 2016).

Above: San Francisquito Creek to the Bay; Below: Sunnyvale East Channel, Calabazas Creek, and San Tomas Aquino Creek to the Bay. (Imagery courtesy of Google Earth)

Creek reconnection to baylands

Historically, channels draining the watersheds surrounding the Bay would deliver freshwater and fine sediment to baylands during flood events, which supported long-term bayland elevation adjustment and maintenance of position in the tidal frame. The rerouting of channels and building of levees along tidal reaches not only caused sedimentation issues at sharp-angle realignments but also cut off adjacent baylands from a regular freshwater and sediment supply while decreasing tidal prism. This caused in-channel sedimentation issues and decreased flood conveyance capacity, which will be exacerbated by SLR. Reconnecting channels to existing and restored baylands could relieve many of these issues and provide a variety of benefits, including:

- Slowing flood velocities, and promoting fine sediment deposition;

- Increasing the long-term delivery of freshwater and sediment to the marsh plain and increasing vertical accretion as sea level rises;

- Increasing tidal prism and channel size, thereby increasing flood conveyance and transport of sediment out to the Bay;

- Improving overall conditions for resident wildlife by establishing salinity gradients and allowing the exchange of nutrients, food resources, energy, and species between the channel and the marsh plain.

Since reconnecting creeks to baylands was identified in the Goals Update (2015) as a key component of bayland resilience under SLR, there has been a regional focus on identifying reconnection opportunities around the Bay. The Flood Control 2.0 effort identified 25 major flood control channels with the potential for creek-bayland reconnection based on the availability of undeveloped land adjacent to the channel that could be restored to tidal marsh, but also showed that there are over 200 channels that currently wind around reclaimed or diked baylands (SFEI-ASC 2017a). As part of the Healthy Watershed Resilient Baylands project, SFEI worked with partners at Valley Water and the South Bay Salt Pond Restoration Project, along with a team of technical advisors, to move forward a recommendation from Flood Control 2.0 and develop a conceptual design for a first-of-its-kind channel-bayland reconnection along the Lower South Bay shoreline (SFEI-ASC 2018) (Figure 4.10). In addition, as part of the 2017-2018 Resilient by Design Bay Area Challenge, the Public Sediment design team worked with Alameda County Flood Control District and other partners to develop a design for lower Alameda Creek that calls for connecting the creek to the adjacent Eden Landing tidal marsh restoration project, thereby providing the marsh with much-needed sediment. These and other efforts can be used as guidance for identifying additional opportunities for creek-bayland reconnection around the region and developing multi-benefit reconnection designs that support long-term bayland resilience while providing flood protection and other vital ecosystem services as sea level continues to rise.

CURRENT CONDITIONS

① Treatment plant discharges wastewater into Bay

② Subsided salt ponds need sediment to recover tidal elevations for restoration and resilience

③ Storage basin in need of upgrade

④ Existing wetlands are at too low an elevation to be sustainable over time with sea-level rise and flooding

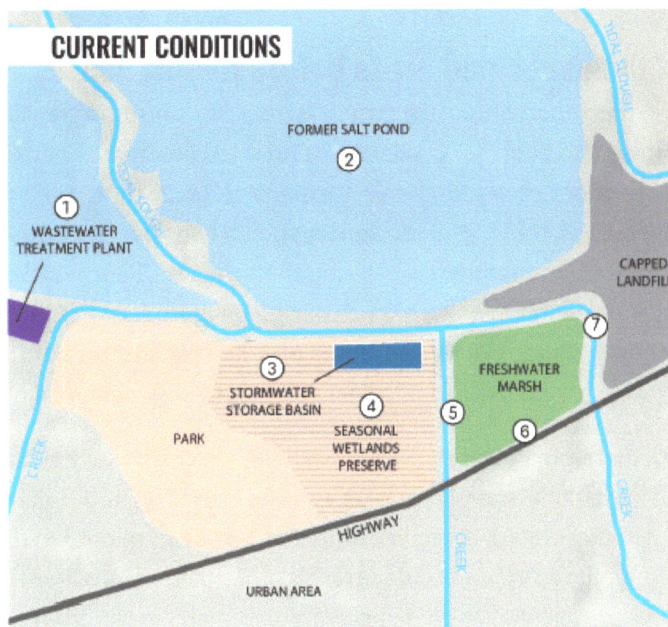

⑤ Levees restrict wildlife movement and cut tidal marsh off from sediment and freshwater exchange, reducing ability of marsh accretion to keep pace with sea-level rise

⑥ Limited wildlife habitat and space for tidal marsh migration as sea levels rise

⑦ Unnatural creek alignment causes sediment accumulation, making creek vulnerable to flooding from sea-level rise and large storms

FUTURE VISION

① Park provides recreation and can accommodate future transition zone habitat as sea levels rise

② Treated wastewater irrigates the ecotone slope to maximize peat accumulation and elevation gain in fresh/brackish marsh

③ Levees lowered around basin to create salt pond habitat within the tidal marsh

④ Tidal marsh area expanded through levee removal and reconnection to fluvial-tidal influence

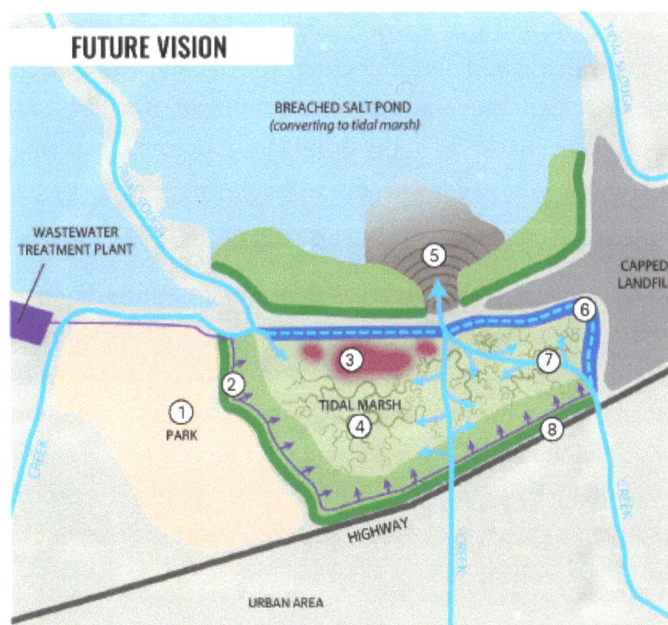

⑤ Realigned creeks deliver sediment into former salt production pond, building elevation conducive to tidal marsh restoration

⑥ Remnant channel provides backwater habitat for fish

⑦ Creeks are reconnected to the marsh plain, driving more rapid elevation gain and long-term sustainability as sea level rises

⑧ Gently sloped levees provide high-tide refugia and habitat for wildlife and space for marsh migration with sea-level rise

Figure 4.10. Vision for connecting Calabazas Creek and San Tomas Aquino Creek to Pond A8 to deliver freshwater and sediment that will support tidal marsh restoration (Lower South Bay) (source: Beller et al. 2019).

Baylands design and management opportunities

Consider filling subsided areas before restoration

The analysis in this report showed a strong increase in long-term resilience when subsided areas were brought to tidal marsh elevations prior to breaching. This is a challenging situation, given that filling prior to breaching is very expensive, takes time, and often is logistically difficult. Nevertheless, restoration designers should consider this option, where sediment is available in sufficient quantities, analyzing the future resilience of the site in question and the local area and subembayment prior to deciding whether or not to fill a subsided area. Restoration sites that require large amounts of sediment and are not filled prior to breaching may cause erosion of tidal flats or otherwise influence sediment availability nearby.

Diked, subsided areas can be filled with inorganic sediment, as has been successfully implemented in Bay tidal marsh restorations like Hamilton and Cullinan Ranch, or they can be filled with natural peat accumulation, as has been piloted in the Delta. Subsidence reversal is being piloted in impounded freshwater marshes on Sherman and Twitchell islands, achieving rapid elevation gains averaging 4 cm/yr (1.6 in/yr) (maximum 7–9 cm/yr (2.8–3.5 in/yr); Miller et al. 2008). These subsidence reversal wetlands or "tule farms" could also be a multi-benefit approach to addressing the subsided areas that currently lie behind levees in the Bay, thus preparing them for future restoration to tidal salt marsh—or simply increasing elevation, providing habitat for wildlife, and reducing levee strain while remaining as impounded freshwater or brackish marsh. Brackish or freshwater marsh plant communities could be sustained in these areas, supported by stormwater, wastewater or creeks (assuming contaminant issues are addressed).

Maximize sediment retention in the baylands

Sediment can be retained in the baylands through natural sediment trapping by plants and centuries-old management techniques, like warping with brushwood fences. Warping is a management process in which water is allowed onto an area and slowed or held in place. As the water velocity drops, the sediment falls out of the water column and accumulates on the surface below. Then the water, now with much less suspended sediment, is able to drain. This is a management-intensive practice that mimics the natural process of turbid waters flooding marsh plains at very high tides, with sediment settling out on the marsh plain during high slack tide (BCDC 2013).

Plants naturally trap sediment in marshes and can be used to enhance accretion. Novel approaches are in development to reintroduce *Suaeda californica* with trellises that allow it to grow taller, providing high-water refuge for marsh wildlife and encouraging greater sediment trapping (Parker and Boyer 2019). This approach could also be applied with pickleweed and *Distichlis spp.* Larger-scale sediment trapping in newly restored areas might be achieved through pre-vegetating with fast-growing marsh plants (that would initially accumulate peat and raise elevations) to increase roughness and sediment trapping at the time of breaching. Similarly, vegetating constructed marsh mounds prior to breaching could create nucleii of wetland plants for clonal spread and sediment trapping.

Above: Hamilton Marsh (Photograph by SFEI); Below: Bay at Coyote Hills (Photograph by Shira Bezalel, SFEI)

Nourish existing tidal flats and marshes as sea levels rise

Sediment can also be placed onto or near existing tidal flats and marshes to augment suspended Bay sediment and sustain natural accretion as the baylands become more supply-limited. Marsh spraying, water column seeding, and shallow water placement are three ideas currently in discussion among restoration practitioners nationally to test strategic placement of dredged sediment (Figure 4.11), but marsh spraying is the only approach that has been piloted to date.

Marsh spraying, also known as thin layer placement, involves "rainbowing" sediment onto a marsh and initially burying vegetation to a depth that is shallow enough that marsh plants can grow through or recolonize (Ray 2007). While research on thin layer placement dates back to the 1970s (Ray 2007), the majority of pilot projects have occurred in marshes located on the East and Gulf coasts, which are characterized by notably different vegetation communities and tidal ranges compared to West Coast marshes. The first thin layer placement sediment augmentation pilot study on an existing marsh in California took place in 2016 at the Seal Beach National Wildlife Refuge (Thorne et al. 2019), and marked an important case study to guide future approaches in West Coast marshes.

Water column seeding is a less-direct sediment placement approach in which dredged sediment is released during a flood tide at the entrance of a marsh channel so the tides can transport the sediment onto the marsh plain. Similarly, shallow water placement involves releasing sediment offshore within the shallow subtidal zone of the Bay to be resuspended and transported by the tides and wind-waves onto nearby tidal flats and marshes. There are no local examples of water column seeding or shallow water placement in San Francisco Bay for marsh nourishment purposes, but lessons learned from case studies in other locations such as the "mud-motor" experiment at the Port of Harlingen in the Netherlands could be used to inform next steps (Baptist et al. 2019).

The ecological trade-offs of each strategic placement technique need to be considered before an approach is

1. MARSH SPRAYING

Dredged sediment is sprayed directly onto the marsh surface, which can increase accretion beyond natural rates.

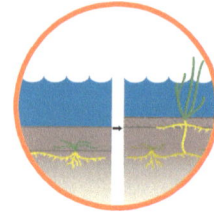

Vegetation is buried with sediment during spraying, affecting habitat quality and quantity for marsh wildlife. New shoots recolonize over time or emerge from buried rhizomes.

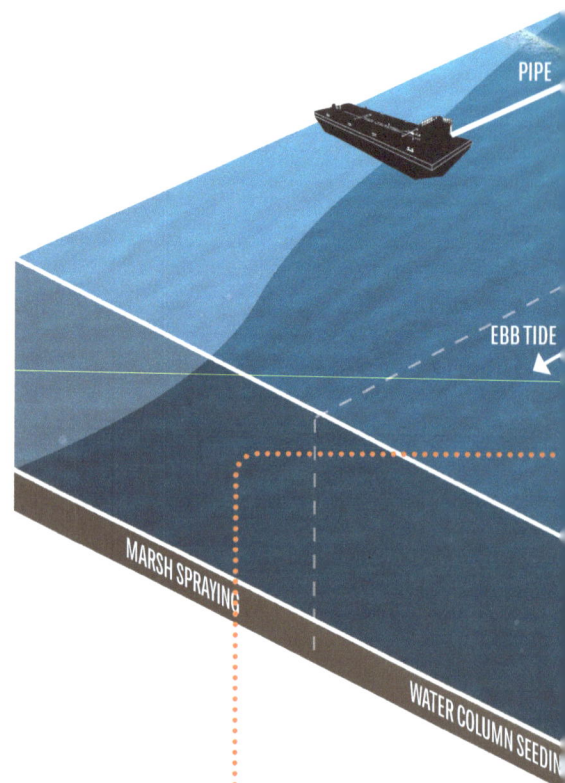

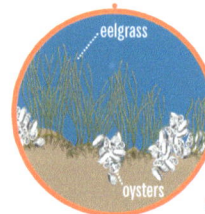

PIPE

EBB TIDE

MARSH SPRAYING

WATER COLUMN SEEDING

eelgrass

oysters

Areas with eelgrasses and oysters should be avoided during shallow water placement.

Figure 4.11. Illustration of three strategic placement methods for dredged sediment. (Illustration by Katie McKnight, SFEI)

2. WATER COLUMN SEEDING

Sediment is released into the water column at the marsh channel entrance during an incoming tide to increase suspended sediment concentrations in the water column.

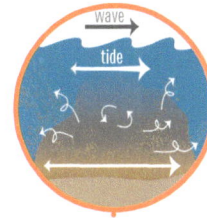

Wave and tidal current energy resuspend the placed sediment and move it primarily landward.

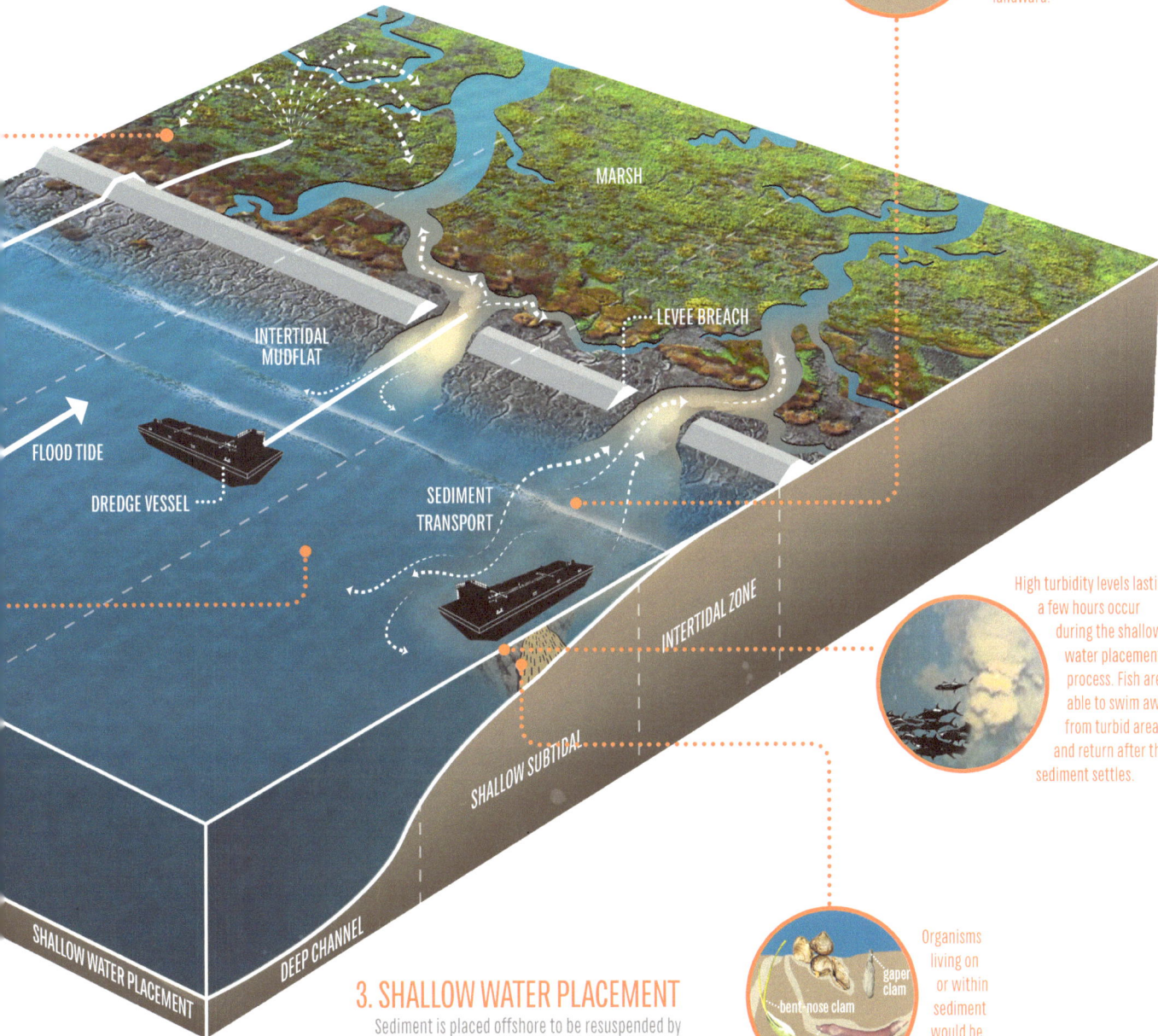

MARSH

LEVEE BREACH

INTERTIDAL MUDFLAT

FLOOD TIDE

DREDGE VESSEL

SEDIMENT TRANSPORT

INTERTIDAL ZONE

High turbidity levels lasting a few hours occur during the shallow water placement process. Fish are able to swim away from turbid areas and return after the sediment settles.

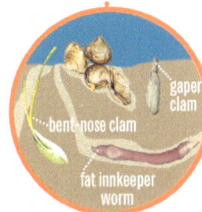

SHALLOW SUBTIDAL

SHALLOW WATER PLACEMENT

DEEP CHANNEL

3. SHALLOW WATER PLACEMENT

Sediment is placed offshore to be resuspended by wave and tide action and then transported by tidal currents onto the marshes.

gaper clam

bent-nose clam

fat innkeeper worm

Organisms living on or within sediment would be buried.

selected. For instance, while marsh spraying initially covers marsh vegetation and the organisms living in the marsh, water column seeding increases the turbidity of the channel and surrounding shallows, which could impact fish and other organisms, and shallow water placement buries organisms in the shallow subtidal zone. Both water column seeding and shallow water placement employ natural processes to transport sediment to baylands habitats, which could have a lower negative impact to the marsh and its resident wildlife but a higher rate of sediment loss.

Manage lateral marsh resilience

Management actions that reduce marsh-edge (or marsh shoreline) erosion and take advantage of adjacent open space at suitable elevations for landward marsh migration will be critical to protecting baylands and maximizing their resilience as sea level rises. The key factors that influence the rate of marsh shoreline retreat or expansion include wind-waves, sediment supply, vegetative structure, and SLR (Bayland Goals 2015). Actions to reduce marsh shoreline erosion should promote rather than inhibit the critical processes that support marsh health—namely the movement of sediment, water, and organisms onto marshes. A prime example of this is the creation of marsh-fringing barrier beaches along eroding marsh shorelines. These features were historically common in the Bay (SFEI and Peter Baye 2020) and can act as important defenses for marshes in appropriate locations by breaking up wave energy along the marsh edge (Barnard et al. 2013). They are also adaptable to changing conditions and can provide ecological benefits when they retreat over salt marshes during storms, creating elevated ridges that support high marsh plants and high water refuge habitat (SFEI and Peter Baye 2020). By contrast, the placement of riprap along marsh shorelines creates a physical barrier that interrupts important ecological flows between tidal flats, marsh, and other subtidal habitats. Riprap or similar hard engineered structures tend to reflect wave energy instead of dissipating it, which could negatively impact adjacent tidal habitats (Gittman et al. 2014). It is important to note, however, that erosion along a marsh's bayward edge or fronting tidal flat may not always be to the detriment of a marsh: in some cases, marsh edge erosion could reintroduce sediment into the water column which could then end up depositing higher on the marsh plain, allowing the marsh to maintain elevation at the expense of preserving the full marsh area (Hopkinson et al. 2018). As novel approaches to stabilizing marsh edge erosion evolve, it will be important to keep in mind the interconnected processes occurring throughout the baylands and potential trade-offs that may arise.

Creating a pathway for marsh systems to migrate upslope will be critical for maintaining baylands if sediment accretion does not keep pace with SLR. As mentioned previously, opportunities for marsh migration in San Francisco Bay are greatly limited due to urbanization. Where feasible, though, protecting and preparing undeveloped areas for baylands to migrate as sea level rises will be an essential component of bayland resilience. Other factors that could influence a marsh's ability to migrate inland with SLR include: the amount of freshwater and mineral sediment inputs to support peat production and marsh colonization; competition by invasive species; and factors that could impact plant dispersal and recruitment including salinity, bulk density, permeability, and soil fertility (Goals Project 2015; SFEI and SPUR 2019). A recent analysis of potentially suitable areas for baylands to migrate inland identified approximately 14,400 acres (~5,800 ha) of undeveloped land between current extreme tide elevation and the elevation that would be inundated with 2 m (6.6 ft) of SLR (SFEI and SPUR 2019) (Figure 4.12). Approximately 70% of this area is unprotected and could be subject to future development (SFEI and SPUR 2019).

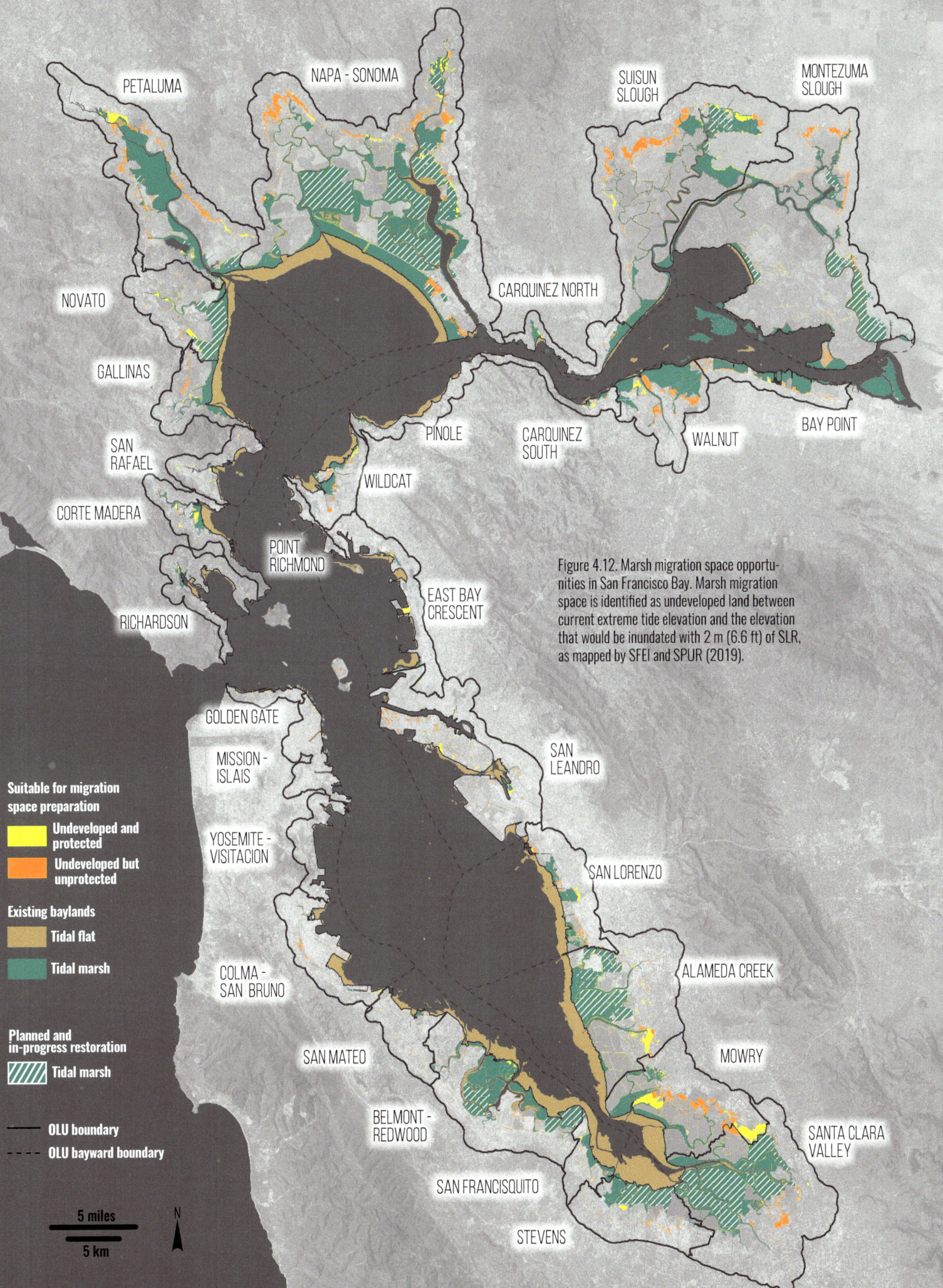

PETALUMA

NAPA - SONOMA

SUISUN
SLOUGH

MONTEZUMA
SLOUGH

NOVATO

CARQUINEZ NORTH

GALLINAS

SAN
RAFAEL

PINOLE

CARQUINEZ
SOUTH

WALNUT

BAY POINT

CORTE MADERA

WILDCAT

POINT
RICHMOND

EAST BAY
CRESCENT

RICHARDSON

Figure 4.12. Marsh migration space opportunities in San Francisco Bay. Marsh migration space is identified as undeveloped land between current extreme tide elevation and the elevation that would be inundated with 2 m (6.6 ft) of SLR, as mapped by SFEI and SPUR (2019).

GOLDEN GATE

MISSION -
ISLAIS

SAN
LEANDRO

YOSEMITE -
VISITACION

SAN LORENZO

**Suitable for migration
space preparation**

◼ **Undeveloped and
protected**

◼ **Undeveloped but
unprotected**

Existing baylands

◼ Tidal flat

◼ Tidal marsh

COLMA -
SAN BRUNO

ALAMEDA CREEK

**Planned and
in-progress restoration**

▨ Tidal marsh

SAN MATEO

MOWRY

BELMONT -
REDWOOD

SANTA CLARA
VALLEY

— OLU boundary

--- OLU bayward boundary

SAN FRANCISQUITO

STEVENS

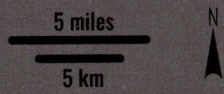

5 miles

5 km

N

Don Edwards Wildlife Refuge (Photo by Abhishek Vanamali, courtesy CC 2.0)

Coordinating Sediment Reuse

There are several efforts in the region aimed at increasing the beneficial reuse of sediment removed from navigation channels, ports and harbors, flood control channels, reservoirs behind dams, construction sites, and other locations for tidal wetland restoration and shoreline adaptation efforts. The longest standing regional effort aimed at beneficial reuse of in-Bay sediment is the Long Term Management Strategy for the Placement of Dredged Material in San Francisco Bay (LTMS) Program. Formed in 1990, the LTMS Program is a collaborative partnership involving regulatory agencies, resources agencies, and stakeholders working together to maximize beneficial reuse of dredged sediment and minimize in-Bay and ocean sediment disposal. The sponsoring agencies include the U.S. Environmental Protection Agency, the U.S. Army Corps of Engineers, the San Francisco Bay Regional Water Quality Control Board, and San Francisco Bay Conservation and Development Commission (BCDC). The LTMS has set the goal of having 40% of in-Bay dredged sediment go to beneficial reuse, 40% go to ocean disposal, and 20% go to in-Bay disposal (LTMS 1998). Since its inception, the LTMS Program has successfully reduced in-Bay disposal of dredged sediment to 1.25 million cubic yards per year (LTMS 2018).

SediMatch, a collaborative program of the San Francisco Bay Joint Venture (SFBJV), San Francisco Estuary Partnership (SFEP), BCDC, SFEI, and others, is a regional sediment reuse effort that is an outgrowth of the LTMS Program. The overall goal of SediMatch is to bring together those that need sediment (e.g., bayland and stream habitat restoration community) with those that have sediment (e.g. flood control agencies, Bay dredging community) to discuss challenges and find mutually beneficial strategies to increase reuse of dredged sediment at habitat restoration sites. One of the main components is an online tool (database and web interface) to match available Bay and watershed sediment with bayland and stream restoration projects and other opportunities for beneficial reuse.

In addition, there are sediment reuse strategies being developed that focus solely on upland sediment sources. For example, Zone 7 Water Agency is working with SFEI, the SFBJV, and other partners to bring together lessons learned from managers and practitioners to develop a Coarse Sediment Reuse Strategy. The Strategy will outline a framework for overcoming reuse challenges and enabling beneficial uses for sediment that is removed from channels and other depositional locations. It will also outline mechanisms for making the reuse beneficial to those removing the sediment, whether it be compensation, ease of permitting, or some other benefit.

Beneficial reuse of dredged sediment will be essential for achieving in-Bay habitat restoration and endangered species goals, as well as building natural infrastructure needed to address coastal flooding (Goals Project 2015). However, the sediment needed to achieve restoration and flood projection goals as sea level continues to rise is very high. To help meet the need, we need to continue to work at both the local and regional scale to increase the amount of dredged sediment that goes towards beneficial reuse.

Updating the Approach to Resource Reuse for Bayland Resilience

Various hurdles inhibit the optimization of resources that have traditionally been considered waste products. Partially, conceptualizing waste resources for their reuse potential is challenging from a cultural perspective, especially in an environmental context where recent restoration efforts have focused on improving natural processes and landscape functions. However, SLR has forced a shift in the restoration community's considerations of dredged sediment, and a heightened emphasis on its use in the regulatory community for bayland and shoreline resilience, for example.

Similar shifts in thinking about the utility of other nontraditional physical resources may also emerge as the impacts of a long-term mass imbalance of regional sediment come into clearer focus. Expanded research into and consideration of these resources, their current management practices' impacts on the environment, and their potential impacts on the baylands under an alternative management regime are all required for optimizing their potential usefulness in achieving regional adaptation goals. Broadly speaking, three major areas of innovation are evident:

Data Tracking, Sharing and Integrating the Industrial Sector

The industrial ecology of the construction sector concerned with mass transport of soils and construction and demolition wastes is poorly understood partially as a function of its privatization. Environmental contractors operating fleets of heavy equipment and hiring numerous subcontractors to haul large volumes of material compete with one another for bids, which does not incentivize disclosure of project portfolio aspects of interest to planners. Moreover, where the materials of interest are easily received by and disposed of at landfills, there may not be incentive to qualify descriptive factors about the material (geotechnical, contaminants, moisture content, etc.).

The basic relationships of environmental contractors and construction operations may not be malleable within the context of a given sector or competitive market, thus impacting the amount of data tracking or data sharing. However, despite these challenges, there are opportunities for greater resource efficiency in local reuse and recycling, reduced environmental impacts associated with long-haul disposal, and satisfying the material needs of projects that rely upon this sector (e.g., shoreline restoration, environmental mitigation, and adaptation-oriented projects). Understanding the relevant needs, limitations and incentives at play within and across sectors should help identify and spotlight the most critical factors of consideration for reuse of various materials, and potentially compel increased cross-domain data sharing for mutually beneficial projects.

Project Coordination and Strategic Planning

While the day-to-day tracking and sharing of information and data related to the industrial and material ecology of certain resources of interest is important to develop more fine-grained understanding for improved reuse, long-term thinking about these resources is a critical concern for regional shoreline restoration and adaptation planning. An expanded consideration of the scope and scale of projects oriented toward multi-decadal coastal climate adaptation is recognized as standard practice in strategic planning, and may reveal or refine rationales for resource management that vary considerably (from an economic, ecologic, or societal perspective) from those compelled by short-term processes driven by local governance.

Just as the long-term timescales required for considering climate adaptation planning are intuitive, adaptation planning that operates at large geographic scopes is also logical. Regional planning is appropriate for various reasons, including the non-local nature of critical infrastructure networks and the free flow of people, products and resources across jurisdictional boundaries that is critical for socioeconomic functionality in metropolitan areas. Just as the region organizes institutions and initiatives based on expected population growth and development trends, improved forecasting and scenario assessments facilitated through industry's participation may reveal major opportunities for cooperation. And without planning that addresses regional resources and shared priorities related to adaptation and sustainability, balkanization and counter-productive governance approaches may arise.

Evolution of Regulatory and Permitting Frameworks

Consideration of resource management and optimization in the context of expanding spatiotemporal planning horizons leads to important questions about the opportunities for innovation. Especially as it pertains to climate adaptation, many complex and daunting challenges are emerging in regions where long-standing regulatory frameworks were designed to manage development and conservation trends. As societies and wildlife habitats become increasingly imperiled by the negative impacts of climate change and SLR, increasingly bold and ambitious actions may need to be taken, or become the overarching theme of adaptation-related planning efforts.

As such, incremental, experimental and small-scale pilot projects that serve as proofs of concept for broader application should be undertaken, focusing on the non-traditional resources described herein. Projects that aim to successfully integrate nontraditional resources into regionally beneficial schemes for shoreline restoration, flood mitigation and adaptation generally should be prioritized for permitting and funding, including granting waivers at the appropriate governmental and regulatory levels for expediting and systematically evaluating projects with major beneficial effects for the region's future prosperity. §

Create a place-based sediment management strategy

The sediment management measures (or actions) described above should rarely be implemented independently. Rather, to address sediment management objectives in watersheds and downstream baylands, measures focused on improving natural sediment processes and beneficially reusing sediment need to be combined to create sediment strategies (Figure 4.13). Strategies should be developed at the OLU scale through a collaborative process that includes a wide range of stakeholders, and should be adaptable over time as conditions change. They need to be built upon the best available science and site-specific information, with the measures included being selected based on an understanding of current and likely future conditions and through an examination of trade-offs. This section details the key steps at play when developing a place-based sediment strategy.

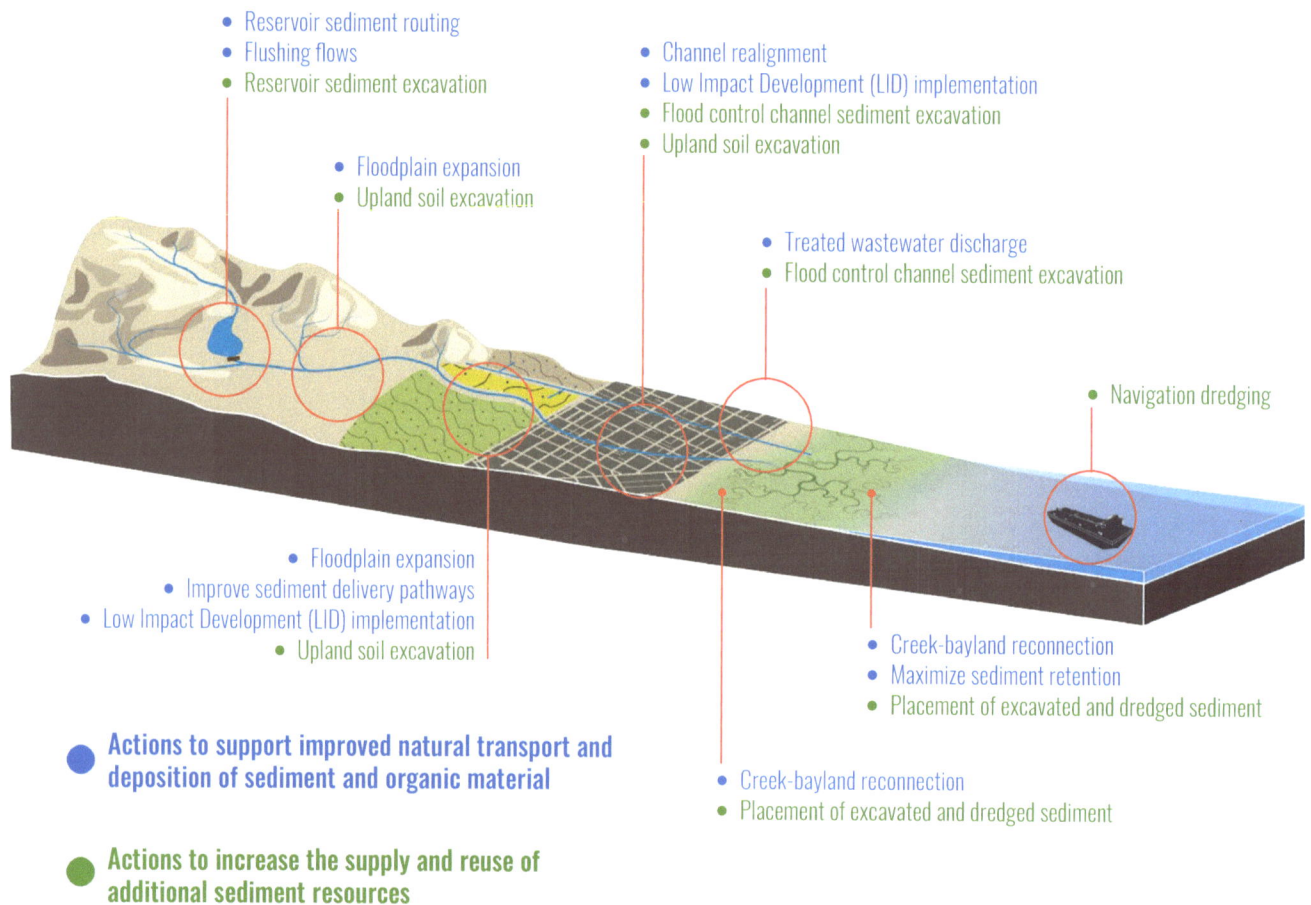

- Reservoir sediment routing
- Flushing flows
- Reservoir sediment excavation

- Channel realignment
- Low Impact Development (LID) implementation
- Flood control channel sediment excavation
- Upland soil excavation

- Floodplain expansion
- Upland soil excavation

- Treated wastewater discharge
- Flood control channel sediment excavation

- Navigation dredging

- Floodplain expansion
- Improve sediment delivery pathways
- Low Impact Development (LID) implementation
- Upland soil excavation

- Creek-bayland reconnection
- Maximize sediment retention
- Placement of excavated and dredged sediment

- Creek-bayland reconnection
- Placement of excavated and dredged sediment

Actions to support improved natural transport and deposition of sediment and organic material

Actions to increase the supply and reuse of additional sediment resources

Figure 4.13. The range of sediment management measures (or actions) to be considered in a sediment management strategy along the gradient from upper watershed to Bay.

Process for creating a place-based sediment management strategy

- *Compile information for local sediment supply and bayland sediment demand and consider range of future climate outcomes*

  The results for the regional analysis done for this study are inherently broad and can be used as a starting point to understand the general trends the future may hold with respect to the comparison of bayland sediment demand and supply. In order to fully understand bayland vulnerability, additional information related to potential bayland sediment demand and sediment supply from both upstream watersheds and the Bay for a broad range of climate futures should be compiled for individual OLUs. For bayland sediment demand, additional SLR projections besides the two considered here should be considered and site-specific data should be used to generate future demand values (e.g., bulk density values for tidal marsh and tidal flat). For bayland sediment supply, the amount available from adjacent watersheds should be determined where possible from calibrated numerical models run with downscaled climate model precipitation data, and the amount available from the Bay should be from numerical modeling of future conditions or from current local suspended sediment values on or near the baylands of interest (e.g., Lacy et al. 2020). In addition, a local value of the fraction of bayland sediment supply available for bayland deposition should be determined by direct measurement, numerical modeling, or from values for baylands with a similar physical setting (e.g., sediment supply, marsh elevation, marsh vegetation type, channel drainage density).

- *Develop a long-term vision for sediment management in watersheds and adjacent baylands that identifies multi-benefit management measures*

  For long-term sediment management for bayland resilience to be successful, coordinated visions should be developed by watershed and adjacent bayland managers within the same OLU. This will allow development of multi-benefit opportunities that support both watershed and bayland ecosystems. Vision development should include local landowners, stakeholders, and regulatory agency representatives, and focus on using the best available data to identify a range of management opportunities that improve current and future sediment supply, transport, and deposition dynamics, while supporting the achievement of established management goals (e.g., TMDL for watershed fine sediment loading). For watersheds, the focus should be on identifying opportunities for improved sediment delivery to channels and improved transport of fine sediment downstream to the baylands in a manner that benefits aquatic habitat. Examples include landscape modifications that re-establish historical sediment delivery pathways, realigning creeks to transport sediment more efficiently, and affecting storm hydrographs to get more fine sediment onto downstream baylands. Opportunities for managing dams to release sediment from impounded reservoirs and beneficially reusing sediment removed from reservoirs, flood control channels, and other sources should also be identified where possible. For baylands, the focus should be on identifying opportunities for transporting watershed sediment directly onto baylands during storm events in a manner that supports sensitive bayland species and increasing the trapping of both watershed and Bay sediment.

- *Assess management options through landscape scenario planning*

  Following Vision development, the identified watershed and bayland management opportunities can be combined into several scenarios to assess key ecosystem trade-offs at the OLU scale. For example, there could be considerable benefits to the bayland ecosystem from increasing watershed fine sediment supply, but the trade-off could be the potential to negatively impact resident fish habitat in the watershed. There could also be a situation where there are considerable benefits to both watershed and bayland ecosystems by coordinated implementation of a few opportunities, but the trade-off could be that the cost is so high that no additional opportunities could be considered. Assessing trade-offs for each scenario should be done quantitatively using a suite of metrics that address the full spectrum of management considerations.

- *Construct a local strategy to implement watershed and bayland sediment management measures over time*

  Following an in-depth landscape scenario planning exercise, those developing the Vision have the information needed to construct a watershed-bayland sediment management strategy that identifies management measures (or actions) that can be implemented over time to improve watershed sediment supply to adjacent baylands. The strategy should lay out the components of each of the measures developed from the opportunities, the anticipated management outcomes, and the recommended order of implementation. The recently released San Francisco Bay Shoreline Adaptation Altas (SFEI and SPUR 2019) provides example strategies for shoreline adaptation under different landscape settings that can provide guidance for developing watershed-bayland sediment management strategies.

Ridgway's Rail at Pt. Isabel (Photo by Becky Matsubara, courtesy CC 2.0)

Fill Critical Knowledge Gaps (Research)

The results from this study and many others provide the science that can be used as the building blocks of sediment strategies that support bayland resilience. However, many knowledge gaps remain with respect to understanding potential future conditions under a changing climate and the factors controlling bayland sediment demand and supply at a range of spatial and temporal scales. Many researchers have recently put forth ideas about the knowledge gaps pertaining to Bay sediment science that need to be addressed to improve management approaches (e.g., BCDC 2016, SFEI-ASC 2017a, Schoellhamer et al. 2018). Here, a list of some of the most high-level critical knowledge gaps specifically related to the assessment of future bayland demand compared to sediment supply is provided. Many of these gaps have already been discussed in previous sections.

Sandbar dredge at Don Edwards (Photo by Charlie Day, courtesy CC 2.0)

Element of Bayland Resilience	Critical Knowledge Gaps
Sediment Demand for Vertical Accretion	Estimates of the sediment needed for Bay channels and shallows to keep pace with SLR
	The rate of in-situ compaction of placed sediments of different grain sizes within tidal marshes, tidal flats, shallows, and channels
	A regional dataset of in-situ dry bulk density and organic matter content for soil cores distributed among and within bayland habitats (including shallows and channels) and across salinity gradients
	A regional dataset of dry bulk density and organic matter content for soil samples from tidal, fluvial, and upland sediments excavated for beneficial reuse
	Updated regional mapping of existing intertidal and subtidal habitats and areas with plans to be or in the process of being restored
	Updated topobathymetric data for polders slated to be breached and/or restored to tidal marsh
Lateral Movement	The impacts of changing climatic conditions and Bay sediment supply on shoreline erosion rates around the Bay
Sediment Supply	Estimates of future Bay sediment supply for a range of climate and land use futures
	The impact of Delta restoration scenarios on future Delta sediment supply to the Bay
	The impacts of changing sea level, precipitation patterns, and Bay sediment supply on the subtidal Bay bed erosion and deposition dynamics
	The impacts of changing sea level, precipitation patterns, and Bay sediment supply on sediment flux at between subembayments and at the Golden Gate
	The impacts of changing precipitation patterns and land use/land cover on watershed flow-sediment load relationships for all Bay tributaries
	Effective approaches for strategic placement of sediment to increase sediment supply to baylands
	The fraction of Bay and watershed sediment that is available for bayland deposition and the controlling factors
Organic Matter Accumulation	The impacts of SLR and changing precipitation patterns and flow to the Bay on marsh salinity and organic matter accumulation rates
	The relationship between organic matter accumulation and treated wastewater discharge for a range of horizontal levee configurations

Monitor to Track Bayland Resilience

Monitoring the conditions of baylands will be essential to assess their ability to keep pace with SLR and the degree to which sediment management actions are supporting long-term resilience. This includes frequent measurements of local sediment processes and topographic change (e.g., annual net sediment flux to a marsh, annual net sediment deposition rates on a marsh, shoreline erosion/progradation rates) and key bayland resilience metrics based on monitoring data (e.g., tidal flooding depth and duration). Currently, bayland monitoring is typically done at the site scale as part of project compliance required by regulatory agencies or by academic institutions and research organizations seeking to understand local processes (e.g., Buffington et al. 2020, Lacy et al. 2020). However, there are new efforts focused on a coordinated regional approach to bayland monitoring that provide recommendations for monitoring actions, locations, and frequency based on previous monitoring approaches and an assessment of what is needed to track bayland resilience over time.

SCIENCE FRAMEWORK
San Francisco Bay Wetlands Regional Monitoring Program

The San Francisco Estuary Wetlands Regional Monitoring Program (WRMP) is being created to provide the scientific understanding of tidal marsh ecosystems needed to conserve them into perpetuity. The WRMP is designed to identify thresholds of marsh ecosystem response to changes in sea level and sediment supply that trigger conservation actions, to test and improve the efficacy of the actions, and to develop models that successfully forecast when and where the actions should happen (WRMP 2020). To develop this capacity, the WRMP must answer the following management questions:

1. Where are the region's tidal marshes and marsh projects, and what net landscape changes in area and condition are occurring?

2. How are external drivers, such as accelerated SLR, development pressure, and changes in runoff and sediment supply, impacting tidal marshes?

3. How do policies, programs, and projects to protect and restore tidal marshes affect the distribution, abundance, and health of plants and animals?

4. What new information is needed to better understand regional lessons from tidal marsh restoration projects in the future?

5. How do policies, programs, and projects to protect and restore tidal marshes benefit and/or impact public health, safety, and recreation?

The WRMP has developed a science framework based on standardized methods to monitor key indicators of tidal marsh abundance, distribution, and condition. The science framework includes a matrix of indicators (i.e., factors and processes to quantify), metrics

Above: Codornices Creek at the Bay; Below: wetlands at Oro Loma (Photographs by Shira Bezalel, SFEI)

(i.e., the exact methods of data collection and quantification), sources of existing data, and data publication and management protocols. The framework also outlines the regional network of permanent long-term monitoring sites, reference sites for projects, and project sites (new and recent restoration or mitigation projects). The WRMP will compare projects to each other and over time, relative to their target conditions, in the context of climate and land use change. The WRMP and the Regional Monitoring Program for Water Quality in San Francisco Bay (Bay RMP) will be closely coordinated with each other and with related research and monitoring efforts throughout the estuary.

SEDIMENT MONITORING AND MODELING STRATEGY
Regional Monitoring Program for Water Quality in San Francisco Bay

In 2016, the Bay RMP created the Sediment Workgroup to provide technical oversight and stakeholder guidance on Bay RMP studies addressing questions about sediment delivery, sediment transport, dredging, and beneficial reuse of sediment within San Francisco Bay. There are five governing Management Questions that Workgroup efforts address:

1. What are acceptable levels of chemicals in sediment for placement in the Bay, baylands, or restoration projects?

2. Are there effects on fish, benthic species, or submerged habitats from dredging or placement of sediment?

3. What are the sources, sinks, pathways, and loadings of sediment and sediment-bound contaminants to and within the Bay and subembayments?

4. How much sediment is passively reaching tidal marshes and restoration projects, and how could the amounts be increased by management actions?

5. What are the concentrations of suspended sediment in the Estuary and its subembayments?

Currently, the Sediment Workgroup is in the process of developing a Sediment Monitoring and Modeling Strategy that addresses Workgroup Management Questions 3–5. The Strategy will also help address WRMP Management Questions 2 and 4. The overarching goal of the Strategy is to provide a planning framework and work plan with monitoring and modeling elements that address information gaps related to sediment delivery to and movement within the Bay. The Strategy includes a prioritized list of actions that will help forward the understanding of the amount of sediment supplied to the Bay from the Delta and Bay tributaries; sediment flux between subembayments and at the Golden Gate; sediment flux between Bay deep Bay, shallows, tidal flats, tidal channels, and tidal marshes; and suspended sediment concentration and settling velocities on tidal flats and tidal marshes.

SEDIMENT MONITORING AND RESEARCH STRATEGY
San Francisco Bay Conservation and Development Commission

In October 2015, BCDC hosted a two-day workshop to identify priority management research needs around sediment processes in the Bay Area and to discuss the formulation of a prioritized scientific research strategy for the region. Participants were from a variety of work sectors (government, research, consulting, management, regulation) and had a wide breadth of expertise (hydrology, geomorphology, flood control, wetland management, dredging, sediment transport). During the workshop, participants developed a list of management questions related to sediment supply and demand for several Bay habitats, including beaches, tributaries, marshes, and subtidal areas, and discussed a research and monitoring framework aimed at collecting key sediment data to address the management questions (BCDC 2016).

Currently, BCDC is developing a San Francisco Bay Sediment Monitoring and Research Strategy to help improve management decisions based in large part on the ideas put forth during the October 2015 workshop. The Strategy will address the current status of sediment supply to various regions of the Bay, as well as future resilience of Bay habitats and shoreline. The Strategy will provide a geographic context for sediment research and monitoring, address the importance of scale in research and monitoring activities, and differentiate transport pathways for coarse and fine grain sediment. It will also provide recommendations for both on-the ground monitoring activities and modeling efforts that should be undertaken to fill prioritized information gaps.

San Francisco Bay at Coyote Hills Regional Park (Photographs by Shira Bezalel, SFEI)

Above: 500 acres of restored tide lands, with elk grazing nearby, October 2020 (Photo by Josh Collins, SFEI)
Below: Petaluma River (photo by Darrel Rhea, courtesy of CC 2.0)

5 Where do we go from here?

This report assesses sediment supply and demand for the entire intertidal zone of San Francisco Bay for a range of possible future conditions. While there is much uncertainty in the analysis, the synthetic, quantified approach taken here using the best available science can help guide planning and policy, quite possibly providing the evidence needed to spur a faster pivot in how we approach sediment management.

How do we reconcile the findings here with our community mandate of restoring tidal marsh to a long-term goal of 100,000 acres (~40,000 ha)? The future tidal marsh area considered in this analysis is 75,000 acres (~30,000 ha), which includes existing marsh, evolving marsh, and lands purchased and planned to be restored to marsh. These findings suggest that, even with only 75,000 acres (~30,000 ha) of tidal marsh (i.e., 25,000 acres (~10,000 ha) less than the long-term goal), the sediment supply necessary for long-term marsh resilience as sea level rises may be lacking in many parts of the region.

Changing our sediment management approach, coordination and policies can change the future trajectories of sediment supply and demand and of marsh resilience, at least for the next several decades. We can choose to restore watershed sediment supply and delivery to baylands, harness the potential of marsh plants to build elevation, discover new solutions with experimentation, and tackle all the options suggested in this report. It is important to remember that there is a lot of uncertainty in this analysis, and we have a lot to learn.

The challenge of restoring and supporting extensive tidal marsh over time means that we need to think carefully about priorities. Given that sediment supplies are finite, how will we choose to allocate sediment among the following priorities?

- Maintain current marshes over time (likely means developing marsh nourishing programs later in the century, which will involve policy, and maybe regulation, changes).

- Restore the thousands of acres that have already been purchased and planned for tidal marsh but not yet restored, and maintain those marshes. This ongoing work will need to tackle the issue of whether or not to fill subsided areas before breaching.

- Restore the ~25,000 acres (~10,000 ha) that have not yet been purchased but are needed to attain the 100,000 acre (~40,000 ha) goal, and maintain those marshes over time.

- Focus on new priorities that require less sediment, such as conserving and creating migration space for tidal marshes, and shifting restoration endpoints to subtidal habitats.

As we are sorting out these larger-scale priorities, we also need to address the following questions related to project prioritization (both marsh restoration projects and those that enhance marsh resilience over time).

- Priority based on projected future resilience—should we allocate resources toward retaining the marshes that are most imperiled or those that are likely to last the longest?

- Priority based on ecosystem functions and services—how should we balance priorities among the following marsh benefits?

    - Supporting endemic, imperiled marsh wildlife (e.g., Ridgway's rail and salt marsh harvest mouse)

    - Protecting shoreline communities and infrastructure from tidal flooding (by attenuating waves and spreading out and slowing high water)

    - Nutrient processing (an under-recognized service marshes have been providing that helps prevent eutrophication impacts, the risk of which is likely to worsen over time)

    - Primary productivity (another under-recognized service marshes have been providing that is critical for supporting aquatic, wetland and terrestrial wildlife, including some fisheries like Dungeness crab)

    - Supporting fish and other aquatic wildlife (not just with food but also with structural habitat)

    - Recreation (the value of time in nature for human health has been highlighted by recent research and by the coronavirus pandemic)

As the baylands management community discusses these issues, we should also keep in mind the longer-view possibility that marshes and mudflats will diminish in width as SLR accelerates, so some resources and planning effort should go into considering how these priorities may shift later in the century. The baylands of the future may be very different in extent, type, and location. How do we conserve enough of the complete marsh ecosystem along the estuarine gradient of salinity and tidal range to ensure the desired functions and services remain? This will require careful management of the complex relationships among SLR, sediment supply, and marsh resilience as well as balancing tradeoffs between space for development and marsh migration.

Figuring out these priorities and achieving alignment across the baylands restoration, regulation and management communities is going to be complicated. Different entities may have different prioritization schemes, requiring regional dialogue and coordination to support rapid progress in restoration. Tools that can quantify projected future resilience and ecosystem functions and services will be very helpful to facilitate decision making. Some type of convening or governance structure seems necessary to avoid conflict, which would lead to inefficient use of resources, slowing down of project implementation in this time of urgency, and reduced baylands ecosystem functions and services compared to aligned decision-making. The WRMP is poised to be such a convening structure, and is actively seeking funding to grow into a robust program.

Fortunately, the community of people focused on the baylands has several-decades-long track record of working together across jurisdictions and agencies with different missions. That investment can pay off now, as we need to pull together to face these challenges. We are evolving our approach in many ways already, but must begin to do so more significantly and rapidly to achieve the goal of healthy baylands that provide desired benefits in perpetuity. The ingenuity of the baylands community should not be underestimated, if we realize the crisis is upon us and rise to meet it.

Ridgway's Rail at Arrowhead Marsh (Photo by Becky Matsubara, courtesy of CC 2.0)

References

33 U.S.C. (United States Code) 1251. 1987. Water Quality Act of 1987. Pub.L. 100-4, February 4, 1987.

Alberti, M. 2008. *Advances in urban ecology: integrating humans and ecological processes in urban ecosystems*. No. 574.5268 A4. New York: Springer.

Allen, J. R. L. 1989. Evolution of salt marsh cliffs in muddy and sandy systems: a qualitative comparison of British west coast estuaries. *Earth Surface Processes and Landforms* 14, no. 1: 85-92.

Atwater, B. F., Hedel, C. W., & Helley, E. J. 1977. *Late Quaternary depositional history, Holocene sea-level changes, and vertical crust movement, southern San Francisco Bay, California* (Vol. 1014). US Govt. Print. Off.

BACWA (Bay Area Clean Water Agencies). 2018. Group Annual Report: Nutrient Watershed Permit Annual Report. Bay Area Clean Water Agencies, Oakland, CA.

Baptist, M. J., Gerkema, T., Van Prooijen, B. C., Van Maren, D. S., Van Regteren, M., Schulz, K., ... & Willemsen, P. 2019. Beneficial use of dredged sediment to enhance salt marsh development by applying a 'Mud Motor'. *Ecological engineering*, *127*, 312-323.

BCDC (Bay Conservation & Development Commission). 2016.The Science of Sediment: Identifying Bay Sediment Science Priorities: Workshop summary report. A report prepared by the San Francisco Bay Conservation and Development Commission (BCDC). https://bcdc.ca.gov/sediment/RSMScienceOfSedimentWorkshop.pdf

BCDC (Bay Conservation & Development Commission). 2017. Central San Francisco Bay Regional Sediment Management Plan. https://www.bcdc.ca.gov/sediment/CentralSFBayRSMPlan.pdf

BCDC (Bay Conservation and Development Commission). 2013. Corte Madera Baylands Conceptual Sea Level Rise Adaptation Strategy. Prepared by the San Francisco Bay Conservation and Development Commission and ESA PWA.

Beagle, J., Salomon, M., Grossinger, R.M., Baumgarten, S., & Askevold, R.A. 2015. Shifting Shores: Marsh Expansion and Retreat in San Pablo Bay. SFEI Contribution No. 751.

Beechie, T. J., Sear, D. A., Olden, J. D., Pess, G. R., Buffington, J. M., Moir, H., ... & Pollock, M. M. 2010. Process-based principles for restoring river ecosystems. *BioScience*, *60*(3), 209-222.

Beller, E. E., Spotswood, E. N., Robinson, A. H., Anderson, M. G., Higgs, E. S., Hobbs, R. J., ... & Grossinger, R. M. 2019. Building ecological resilience in highly modified landscapes. *BioScience*, *69*(1), 80-92.

Beyeler, M., Mehaffy, M., Connor, M., Doehring, C., Lowe, J., Grossinger, R., Senn, D., Novick, E. 2015. Decentralized Wastewater Discharges and Multiple Benefit Natural Infrastructure: Preliminary Analysis and Next Steps (Final Project Report). East Bay Dischargers Authority.

Bobylev, N. 2016. Underground space as an urban indicator: measuring use of subsurface. *Tunnelling and Underground Space Technology*, *55*, 40-51.

Bouma, T. J., Van Belzen, J., Balke, T., Van Dalen, J., Klaassen, P., Hartog, A. M., ... & Herman, P. M. J. 2016. Short term mudflat dynamics drive long term cyclic salt marsh dynamics. *Limnology and Oceanography*, *61*(6), 2261-2275.

Brin, L. D., Valiela, I., Goehringer, D., & Howes, B. 2010. Nitrogen interception and export by experimental salt marsh plots exposed to chronic nutrient addition. *Marine Ecology Progress Series*, *400*, 3-17.

Brinson, M. M., Christian, R. R., & Blum, L. K. 1995. Multiple states in the sea-level induced transition from terrestrial forest to estuary. *Estuaries, 18*(4), 648-659.

Buffington, K. J., Janousek, C. N., Thorne, K. M., & Dugger, B. D. 2020. Spatiotemporal Patterns of Mineral and Organic Matter Deposition Across Two San Francisco Bay-Delta Tidal Marshes. *Wetlands*, 1-13.

Byrne, R., Ingram, B. L., Starratt, S., Malamud-Roam, F., Collins, J. N., & Conrad, M. E. 2001. Carbon-isotope, diatom, and pollen evidence for late Holocene salinity change in a brackish marsh in the San Francisco Estuary. *Quaternary Research, 55*(1), 66-76.

Caffrey, J. M. 1995. Spatial and seasonal patterns in sediment nitrogen remineralization and ammonium concentrations in San Francisco Bay, California. *Estuaries, 18*(1), 219-233.**995**

Callaway, J. C., Nyman, J. A., & DeLaune, R. D. 1996. Sediment accretion in coastal wetlands: a review and a simulation model of processes. *Current topics in wetland biogeochemistry, 2*, 2-23.

Callaway, J. C., Borgnis, E. L., Turner, R. E., & Milan, C. S. 2012. Carbon sequestration and sediment accretion in San Francisco Bay tidal wetlands. *Estuaries and Coasts, 35*(5), 1163-1181.

Cecchetti, A. R., Stiegler, A. N., Graham, K. E., & Sedlak, D. L. 2020. The horizontal levee: a multi-benefit nature-based treatment system that improves water quality and protects coastal levees from the effects of sea level rise. *Water Research X*, 100052.

Cohen, A. N. 2008. Sources and Impacts of Sediment Inputs into the Water Column of San Francisco Bay. A Report for the Subtidal Goals Project of the California Coastal Conservancy, National Oceanic and Atmospheric Administration, San Francisco Bay Conservation and Development Commission, and Association of Bay Area Governments. San Francisco Estuary Institute, Oakland, CA. http://bioinvasions.org/wp-content/uploads/2008-Subtidal-GoalsSediment-Inputs.pdf

Collins, J. N.; Brewster, E.; Grossinger, R. M. 1999. *Conceptual models of freshwater influences on tidal marsh form and function, with an historical perspective.* SFEI Contribution No. 327. Department of Environmental Services: City of San Jose, CA. p 237 pp.

Conomos, T. J., Smith, R. E., & Gartner, J. W. 1985. Environmental setting of San Francisco Bay. In *Temporal dynamics of an estuary: San Francisco Bay* (pp. 1-12). Springer, Dordrecht.

Culberson, S. D., Foin, T. C., & Collins, J. N. 2004. The role of sedimentation in estuarine marsh development within the San Francisco Estuary, California, USA. *Journal of Coastal Research, 20*(4 (204)), 970-979.

California Wetlands Monitoring Workgroup (CWMW). 2020. EcoAtlas. https://www.ecoatlas.org

Davis, J., Fono, L., Williams, D., Johnson, B., Schlipf, R., and Hall, T. 2019. Municipal Wastewater, in The Pulse of the Bay: Pollutant Pathways. SFEI Contribution #954. San Francisco Estuary Institute, Richmond, CA.

Delta Modeling Associates. 2015. Analysis of the Effect of Project Depth, Water Year Type and Advanced Maintenance Dredging on Shoaling Rates in the Oakland Harbor Navigation Channel, Central San Francisco Bay 3-D Sediment Transport Modeling Study, Final Report, Prepared for U.S. Army Corps of Engineers, San Francisco District, March 2015.

Deverel, S.J., Drexler, J.Z. , Ingrum, T., & Hart, C.. 2008. Simulated Holocene, recent, and future accretion in channel marsh islands and impounded marshes for subsidence mitigation, Sacramento– San Joaquin Delta. California, USA: REPEAT Project Final Report to the CALFED Science Program of the Resources Agency of California. 60 pp.

Duvall, M. S., Wiberg, P. L., & Kirwan, M. L. 2019. Controls on sediment suspension, flux, and marsh deposition near a bay-marsh boundary. *Estuaries and coasts, 42*(2), 403-424.

Douglas, I., & Lawson, N. 2000. The human dimensions of geomorphological work in Britain. *Journal of Industrial Ecology*, *4*(2), 9-33.

Drexler, J. Z., de Fontaine, C. S., & Brown, T. A. 2009. Peat accretion histories during the past 6,000 years in marshes of the Sacramento–San Joaquin Delta, CA, USA. *Estuaries and Coasts*, *32*(5), 871-892.

DSOD (Division of Safety of Dams). 2020. https://data.cnra.ca.gov/dataset/california-jurisdictional-dams

Duvall, M. S., Wiberg, P. L., & Kirwan, M. L. 2019. Controls on sediment suspension, flux, and marsh deposition near a bay-marsh boundary. *Estuaries and coasts*, *42*(2), 403-424.

Erikson, L. H., Wright, S. A., Elias, E., Hanes, D. M., Schoellhamer, D. H., & Largier, J. 2013. The use of modeling and suspended sediment concentration measurements for quantifying net suspended sediment transport through a large tidally dominated inlet. *Marine Geology*, *345*, 96-112.

Fagherazzi, S. 2013. The ephemeral life of a salt marsh. *Geology*, *41*(8), 943-944.

Fischel, M., & Robilliard, G. A. 1991. Natural Resource Damage Assessment of the Shell Oil Spill at Martinez, California. In *International Oil Spill Conference* (Vol. 1991, No. 1, pp. 371-376). American Petroleum Institute.

Flemming, B. W., & Delafontaine, M. T. 2016. Mass Physical Sediment Properties. In Encyclopedia of Earth Sciences Series. https://doi.org/10.1007/978-94-017-8801-4_350.

Flick, R. E., Murray, J. F., & Ewing, L. C. 1999. *Trends in US tidal datum statistics and tide range: A data report atlas*. Center for Coastal Studies, Scripps Institution of Oceanography.

Flint, L. E., & Flint, A. L. 2012. Downscaling future climate scenarios to fine scales for hydrologic and ecological modeling and analysis. *Ecological Processes*, *1*(1), 2.

Flint, L.E., Flint, A.L., Thorne, J.H., & Boynton, R. 2013. Fine-scale hydrologic modeling for regional landscape applications: the California Basin Characterization Model development and performance. *Ecological Processes*, 2:25.

Flint, L.E., Flint, A.L., & Stern, M.A. 2020. The Basin Characterization Model -- A Regional Water Balance Code: U.S. Geological Survey Techniques and Methods 2020, 112 p.

Foley, M., Christian, B., Goeden, B., Ross, B., Sui, J., & Gravenmier, J. 2019. Dredging and Dredged Sediment Disposal. In The Pulse of the Bay: Pollutant Pathways. SFEI Contribution #954. San Francisco Estuary Institute, Richmond, CA.

Foster-Martinez, M. R., & Variano, E. A. 2018. Biosolids as a marsh restoration amendment. *Ecological Engineering*, *117*, 165-173.

Francalanci, S., Bendoni, M., Rinaldi, M., & Solari, L. 2013. Ecomorphodynamic evolution of salt marshes: Experimental observations of bank retreat processes. Geomorphology, 195:53-65.

Ganju, N. K., & Schoellhamer, D. H. 2006. Annual sediment flux estimates in a tidal strait using surrogate measurements. *Estuarine, Coastal and Shelf Science*, *69*(1-2), 165-178.

Gittman, R. K., Popowich, A. M., Bruno, J. F., & Peterson, C. H. 2014. Marshes with and without sills protect estuarine shorelines from erosion better than bulkheads during a Category 1 hurricane. *Ocean & Coastal Management*, *102*, 94-102.

Goals Project. 1999. Baylands Ecosystem Habitat Goals. A Report of Habitat Recommendations Prepared by the San Francisco Bay Area Wetlands Ecosystem Goals Project. U.S. Environmental Protection Agency and S.F. Bay Regional Water Quality Control Board, San Francisco and Oakland, CA.

Goals Project. 2015. The Baylands and Climate Change: What We Can Do. The 2015 Science Update to the Baylands Ecosystem Habitat Goals Prepared by the San Francisco Bay Area Wetlands Ecosystem Goals Project. California State Coastal Conservancy, Oakland, CA.

Gore, J. A. 2018. *Alternatives in regulated river management*. CRC Press.

Grant, G. E., Schmidt, J. C., & Lewis, S. L. 2003. A geological framework for interpreting downstream effects of dams on rivers. *Water Science and Application*, *7*, 209-225.

Hart, D. D., Johnson, T. E., Bushaw-Newton, K. L., Horwitz, R. J., Bednarek, A. T., Charles, D. F., ... & Velinsky, D. J. 2002. Dam removal: challenges and opportunities for ecological research and river restoration: we develop a risk assessment framework for understanding how potential responses to dam removal vary with dam and watershed characteristics, which can lead to more effective use of this restoration method. *BioScience*, *52*(8), 669-682.

Hopkinson, C. S., Morris, J. T., Fagherazzi, S., Wollheim, W. M., & Raymond, P. A. 2018. Lateral marsh edge erosion as a source of sediments for vertical marsh accretion. Journal of Geophysical Research: Biogeosciences, 123, 2444–2465.

Hu, M., Van Der Voet, E., & Huppes, G. 2010. Dynamic material flow analysis for strategic construction and demolition waste management in Beijing. *Journal of Industrial Ecology*, *14*(3), 440-456.

Inui, T., Yasutaka, T., Endo, K. and Katsumi, T., 2012. Geo-environmental issues induced by the 2011 off the Pacific Coast of Tohoku Earthquake and tsunami. Soils and Foundations, 52(5), pp.856-871.

Jaffe, B., Foxgrover, A., & Finlayson, D. 2011. Mudflat Loss During South San Francisco Bay Salt Pond Restoration - Regional and Global Perspectives on Initial Post-Restoration Changes. South Bay Science Symposium. February 3, 2011. Menlo Park., CA.

Knowles, N. 2010. Potential inundation due to rising sea levels in the San Francisco Bay region. *San Francisco Estuary and Watershed Science*, *8*(1).

Kondolf, G. M., Cada, G. F., & Sale, M. J. 1987. Assessing flushing-flow requirements for brown trout spawning gravels in steep streams. *JAWRA Journal of the American Water Resources Association*, *23*(5), 927-935.

Kondolf, G. M., & Wilcock, P. R. 1996. The flushing flow problem: defining and evaluating objectives. *Water Resources Research*, *32*(8), 2589-2599.

Kondolf, G. M., Gao, Y., Annandale, G. W., Morris, G. L., Jiang, E., Zhang, J., ... & Hotchkiss, R. 2014. Sustainable sediment management in reservoirs and regulated rivers: Experiences from five continents. *Earth's Future*, *2*(5), 256-280.

Kopp, R. E., Horton, R. M., Little, C. M., Mitrovica, J. X., Oppenheimer, M., Rasmussen, D. J., ... & Tebaldi, C. 2014. Probabilistic 21st and 22nd century sea level projections at a global network of tide gauge sites. *Earth's future*, *2*(8), 383-406.

Krone, R. B. 1987. A method for simulating historic marsh elevations. *Journal of Coastal Sediments, 1*: 316-323.

Lacy, J. R., Foster Martinez, M. R., Allen, R. M., Ferner, M. C., & Callaway, J. C. 2020. Seasonal Variation in Sediment Delivery Across the Bay Marsh Interface of an Estuarine Salt Marsh. *Journal of Geophysical Research: Oceans*, *125*(1), e2019JC015268.

Lionberger, M. A., & Schoellhamer, D.H. 2009. A Tidally Averaged Sediment-Transport Model for San Francisco Bay, California. US Geological Survey.

Livsey, D. N., Downing-Kunz, M. A., Schoellhamer, D. H., & Manning, A. J. 2020. Suspended Sediment Flux in the San Francisco Estuary: Part I—Changes in the Vertical Distribution of Suspended Sediment and Bias in Estuarine Sediment Flux Measurements. *Estuaries and Coasts*, 1-17.

LTMS (Long-term Management Strategy). 1998. Long-term Management Strategy for the Placement of Dredged Material in the San Francisco Bay Region. Final Policy Environmental Impact Statement, Programmatic Environmental Impact Report. Volume III. Prepared by the US Army Corps of Engineers, US Environmental Protection Agency, Bay Conservation and Development Commission, San Francisco Bay Regional Water Quality Control Board, and State Water Resources Control Board. https://www.spn.usace.army.mil/Missions/Dredging-Work-Permits/LTMS/October-1998-Volume-3/

LTMS (Long Term Management Strategy). 2011. Dredged Material Management Office (DMMO) Report of Dredging and Placement of Dredged Material in San Francisco Bay in 2010. Annual Report. Prepared by the US Army Corps of Engineers, US Environmental Protection Agency, Bay Conservation and Development Commission, and San Francisco Bay Regional Water Quality Control Board. San Francisco, CA. https://www.spn.usace.army.mil/Missions/Dredging-Work-Permits/Dredged-Material-Management-Office-DMMO/Annual-Reports/

LTMS (Long Term Management Strategy). 2012. Dredging Material and Management Office (DMMO) Report of Dredging and Placement of Dredged Material in San Francisco Bay in 2011. Annual Report. Prepared by the US Army Corps of Engineers, US Environmental Protection Agency, Bay Conservation and Development Commission, and San Francisco Bay Regional Water Quality Control Board. San Francisco, CA. https://www.spn.usace.army.mil/Missions/Dredging-Work-Permits/Dredged-Material-Management-Office-DMMO/Annual-Reports/

LTMS (Long Term Management Strategy). 2013. Dredged Material Management Office (DMMO) Dredging and Placement of Dredged Material in San Francisco Bay January-December 2012 Report. Annual Report. Prepared by the US Army Corps of Engineers, US Environmental Protection Agency, Bay Conservation and Development Commission, and San Francisco Bay Regional Water Quality Control Board. San Francisco, CA. https://www.spn.usace.army.mil/Missions/Dredging-Work-Permits/Dredged-Material-Management-Office-DMMO/Annual-Reports/

LTMS (Long Term Management Strategy). 2014. Dredged Material Management Office (DMMO) Dredging and Placement of Dredged Material in San Francisco Bay January-December 2013 Report. Annual Report. Prepared by the US Army Corps of Engineers, US Environmental Protection Agency, Bay Conservation and Development Commission, and San Francisco Bay Regional Water Quality Control Board. San Francisco, CA. https://www.spn.usace.army.mil/Missions/Dredging-Work-Permits/Dredged-Material-Management-Office-DMMO/Annual-Reports/

LTMS (Long Term Management Strategy). 2015. Dredged Material Management Office (DMMO) Dredging and Placement of Dredged Material in San Francisco Bay January-December 2014 Report. Annual Report. Prepared by the US Army Corps of Engineers, US Environmental Protection Agency, Bay Conservation and Development Commission, and San Francisco Bay Regional Water Quality Control Board. San Francisco, CA. https://www.spn.usace.army.mil/Missions/Dredging-Work-Permits/Dredged-Material-Management-Office-DMMO/Annual-Reports/

LTMS (Long Term Management Strategy). 2016. Dredged Material Management Office (DMMO) Dredging and Placement of Dredged Material in San Francisco Bay January-December 2015 Report. Annual Report. Prepared by the US Army Corps of Engineers, US Environmental Protection Agency, Bay Conservation and Development Commission, and San Francisco Bay Regional Water Quality Control Board. San Francisco, CA. https://www.spn.usace.army.mil/Missions/Dredging-Work-Permits/Dredged-Material-Management-Office-DMMO/Annual-Reports/

LTMS (Long Term Management Strategy). 2017. Dredging Material Management Office (DMMO) Dredging and Placement of Dredged Material in San Francisco Bay January-December 2016 Report. Annual Report. Prepared by the US Army Corps of Engineers, US Environmental Protection Agency, Bay Conservation and Development Commission, and San Francisco Bay Regional Water Quality Control Board. San Francisco, CA. https://www.spn.usace.army.mil/Missions/Dredging-Work-Permits/Dredged-Material-Management-Office-DMMO/Annual-Reports/

LTMS (Long Term Management Strategy). 2018. Long-Term Management Strategy for the Placement of Dredged Sediment in the San Francisco Bay Region: Beneficial Reuse Fact Sheet. May 2018.

LTMS (Long Term Management Strategy). 2019. Dredging and Placement of Dredged Material in San Francisco Bay January-December 2018 Report. Annual Report. Prepared by the US Army Corps of Engineers, US Environmental Protection Agency, Bay Conservation and Development Commission, and San Francisco Bay Regional Water Quality Control Board. San Francisco, CA.

Lu, Q., He, Z. L., & Stoffella, P. J. 2012. Land application of biosolids in the USA: A review. *Applied and Environmental Soil Science*, *2012*.

Magilligan, F. J., & Nislow, K. H. 2005. Changes in hydrologic regime by dams. *Geomorphology*, *71*(1-2), 61-78.

Magnusson, S., Lundberg, K., Svedberg, B., & Knutsson, S. 2015. Sustainable management of excavated soil and rock in urban areas—a literature review. *Journal of Cleaner Production*, *93*, 18-25.

Magnusson, S., Johansson, M., Frosth, S., & Lundberg, K. 2019. Coordinating soil and rock material in urban construction—Scenario analysis of material flows and greenhouse gas emissions. *Journal of Cleaner Production*, *241*, 118236.

Marani, M., d'Alpaos, A., Lanzoni, S., & Santalucia, M. 2011. Understanding and predicting wave erosion of marsh edges. Geophysical Research Letters,38(21).

McKnight, K., Lowe, J. and Plane, E., 2020. Special Study on Bulk Density. A report prepared for the Regional Monitoring Program for Water Quality in San Francisco Bay (RMP). SFEI Contribution 975. San Francisco Estuary Institute, Richmond, CA.

Miles, S. R., & Goudey, C. B. 1997. Ecological subregions of California: section and subsection descriptions. R5-EM-TP-005. San Francisco, CA: US Department of Agriculture, Forest Service, Pacific Southwest Region.

Miller, R. L., Fram, M., Fujii, R., & Wheeler, G. 2008. Subsidence reversal in a re-established wetland in the Sacramento-San Joaquin Delta, California, USA. *San Francisco Estuary and Watershed Science*, *6*(3).

Milligan, B., Holmes, R., Wirth, G., Maly, T., Burkholder, S., & Holzman, J. Dredge Research Collaborative. 2016. DredgeFest California: Key Findings and Recommendations. Retrieved from http://dredgeresearchcollaborative.org/works/dredgefest-california-white-paper/.

Minear, J. T., & Kondolf, G. M. 2009. Estimating reservoir sedimentation rates at large spatial and temporal scales: A case study of California. *Water Resources Research*, *45*(12).

Moffatt & Nichol. 1997. Inventory of San Francisco Bay Area Dredging Projects, Dredged Material Reuse Study. Moffatt & Nichol Engineers, Oakland, CA. http://50.62.26.103/planning/reports/InventoryOfSFBayAreaDredgingProjectsDredgedMaterialReuseStudy_May1997.pdf

Morris, J. T., Sundareshwar, P. V., Nietch, C. T., Kjerfve, B., & Cahoon, D. R. 2002. Responses of coastal wetlands to rising sea level. *Ecology*, *83*(10), 2869-2877.

Morris, G.L., & Fan, J. 1998. Reservoir Sedimentation Handbook. McGraw-Hill, New York; 898 pp.

NHC (Northwest Hydraulic Consultants). 2004. San Francisquito Creek Watershed Analysis and Sediment Reduction Plan Final Report. Prepared for the San Francisquito Creek Joint Powers Authority.

OPC (California Ocean Protection Council). 2018. State of California Sea-Level Rise Guidance 2018 Update. California Ocean Protection Council, California Natural Resources Agency.

Orr, M., Crooks, S., & Williams, P. B. 2003. Will restored tidal marshes be sustainable?. *San Francisco Estuary and Watershed Science*, *1*(1).

Parker, V. T., & Boyer, K. E. 2017. Sea-level rise and climate change impacts on an urbanized Pacific Coast estuary. *Wetlands*, *39*(6), 1219-1232.

Patrick Jr, W. H., & DeLaune, R. D. 1990. Subsidence, accretion, and sea level rise in south San Francisco Bay marshes. *Limnology and Oceanography*, *35*(6), 1389-1395.

Perry, H., Lyndon, A., Soumoy, P., & Goeden, B. 2015. San Francisco Bay sediment: Challenges and opportunities. Poster presented to the 12th Biennial State of the San Francisco Estuary Conference - Sept. 17-18, 2015, Oakland Marriott City Center, Oakland, CA.

Pestrong, R. 1965. The development of drainage patterns on tidal marshes. A technical report under the Office of Naval Research, Contract Nour-4430 (00). *Stanford University Publications, Geological Sciences, 10(2)*.

Pierce, D. W., Kalansky, J. F., & Cayan, D. R. 2018. Climate, Drought, and Sea Level Rise Scenarios for California's Fourth Climate Change Assessment. *Scripps Institution of Oceanography, California Energy Commission*.

Price, S. J., Ford, J. R., Cooper, A. H., & Neal, C. 2011. Humans as major geological and geomorphological agents in the Anthropocene: the significance of artificial ground in Great Britain. *Philosophical Transactions of the Royal Society A: Mathematical, Physical and Engineering Sciences*, *369*(1938), 1056-1084.

Public Sediment. 2019. Volume II: Public Sediment for Alameda Creek. *Resilient By Design: Bay Area Challenge*.

Porterfield, G. 1980. *Sediment transport of streams tributary to San Francisco, San Pablo, and Suisun Bays, California, 1909-66* (Vol. 80, No. 64). US Geological Survey, Water Resources Division.

Ray G.L. 2007. Thin layer disposal of dredged material on marshes: a review of the technical and scientific literature. In ERDC/EL technical notes collection (ERDC/EL TN-07-01). Vicksburg: US Army Engineer Research and Development Center.

Richter, B. D., & Thomas, G. A. 2007. Restoring environmental flows by modifying dam operations. *Ecology and society*, *12*(1).

Roni, P., & Beechie, T. (Eds.). 2012. *Stream and watershed restoration: a guide to restoring riverine processes and habitats*. John Wiley & Sons.

Russo, T. A., Fisher, A. T., & Winslow, D. M. 2013. Regional and local increases in storm intensity in the San Francisco Bay Area, USA, between 1890 and 2010. *Journal of Geophysical Research: Atmospheres*, *118*(8), 3392-3401.

Schile, L. M. 2012. *Tidal wetland vegetation in the San Francisco Bay Estuary: modeling species distributions with sea-level rise* (Doctoral dissertation, UC Berkeley).

Schile, L. M., Callaway, J. C., Morris, J. T., Stralberg, D., Parker, V. T., & Kelly, M. 2014. Modeling tidal marsh distribution with sea-level rise: Evaluating the role of vegetation, sediment, and upland habitat in marsh resiliency. *PloS one*, *9*(2), e88760.

Schoellhamer, D. H. 2011. Sudden clearing of estuarine waters upon crossing the threshold from transport to supply regulation of sediment transport as an erodible sediment pool is depleted: San Francisco Bay, 1999. *Estuaries and Coasts*, *34*(5), 885-899.

Schoellhamer, D.H., Lionberger, M.A., Jaffe, B.E., Ganju, N.K., Wright, S.A., & Shellenbarger, G.G. 2005. *Bay Sediment Budgets: Sediment Accounting 101*. In The Pulse of the Estuary: Monitoring and Managing Water Quality in the San Francisco Estuary. SFEI Contribution 411. San Francisco Estuary Institute, Oakland, CA.

Schoellhamer, D., Marineau, M., McKee, L., Pearce, S., Kauhanen, P., Salomon, M., Dusterhoff, S., Grenier, L., Trowbridge, P. 2018. *Sediment Supply to San Francisco Bay, Water Years 1995 through 2016: Data, Trends, and Monitoring Recommendations to Support Decisions About Water Quality, Tidal Wetlands, and Resilience to Sea Level Rise.* San Francisco Estuary Institute, Richmond, CA.

Schuerch, M., Spencer, T., & Evans, B. 2019. Coupling between tidal mudflats and salt marshes affects marsh morphology. *Marine Geology*, *412*, 95-106.

Schwimmer, R. A. 2001. Rates and processes of marsh shoreline erosion in Rehoboth Bay, Delaware, USA. *Journal of Coastal Research*, 672-683.

SCVWD (Santa Clara Valley Water District). 2005. Baylands Management Unit, Chapter 6 in the Guadalupe Watershed Stewardship Plan. 105 pages. Available at http:// www.valleywater.org/_wmi/Stewardship_plan/Comments/watershedprod.cfm?watershedid=1

SediMatch. 2020. https://sedimatch.sfei.org/.

Serrana, J. M., Yaegashi, S., Kondoh, S., Li, B., Robinson, C. T., & Watanabe, K. 2018. Ecological influence of sediment bypass tunnels on macroinvertebrates in dam-fragmented rivers by DNA metabarcoding. *Scientific reports*, *8*(1), 1-10.

Shirzaei, M., & Bürgmann, R. 2018. Global climate change and local land subsidence exacerbate inundation risk to the San Francisco Bay Area. *Science advances*, *4*(3), eaap9234.

SFEI (San Francisco Estuary Institute). 2012. Napa River Watershed Profile: Past and Present Characteristics with Implications for Future Management for the Changing Napa River Valley. SFEI Contribution #615. Richmond, CA.

SFEI (San Francisco Estuary Institute). 2016. San Francisco Bay Shore Inventory: Mapping for Sea Level Rise Planning. San Francisco Estuary Institute-Aquatic Science Center, Richmond, CA.

SFEI-ASC (San Francisco Estuary Institute-Aquatic Science Center). 2016. San Francisquito Creek Baylands Landscape Change Metrics Analysis. A Report of SFEI-ASC's Resilient Landscapes Program. SFEI Contribution #784. San Francisco Estuary Institute-Aquatic Science Center: Richmond, CA.

SFEI-ASC (San Francisco Estuary Institute-Aquatic Science Center). 2017a. Changing Channels: Regional Information for Developing Multi-Benefit Flood Control Channels at the Bay Interface. San Francisco Estuary Institute-Aquatic Science Center: Richmond, CA.

SFEI-ASC (San Francisco Estuary Institute-Aquatic Science Center). 2017b. Bay Area Aquatic Resource Inventory (BAARI) Version 2.1 GIS Data. San Francisco Estuary Institute-Aquatic Science Center: Richmond, CA.

SFEI-ASC (San Francisco Estuary Institute-Aquatic Science Center). 2018. Resilient Landscape Vision for the Calabazas Creek, San Tomas Aquino Creek, and Pond A8 Area: Bayland-Creek Reconnection Opportunities. A SFEI-ASC Resilient Landscape Program report developed in cooperation with the Healthy Watersheds, Resilient Baylands Design Advisory Team, Santa Clara Valley Water District, and South Bay Salt Ponds Restoration Project, SFEI Contribution #870. San Francisco Estuary Institute-Aquatic Science Center, Richmond, CA.

SFEI and Peter Baye. 2020. New Life for Eroding Shorelines: Beach and Marsh Edge Change in the San Francisco Estuary. SFEI Contribution #984. San Francisco Estuary Institute: Richmond, CA.

SFEI and SPUR. 2019. San Francisco Bay Shoreline Adaptation Atlas: Working with Nature to Plan for Sea Level Rise Using Operational Landscape Units. SFEI Contribution #915. San Francisco Estuary Institute: Richmond, CA.

SFEP (San Francisco Estuary Partnership). 1994. Comprehensive conservation and management plan. Prepared by the Association of Bay Area Governments for the U.S. EPA. Oakland, CA. https://www.sfestuary.org/wp-content/uploads/2018/01/1993-CCMP.pdf

SFEP (San Francisco Estuary Partnership). 2016. Comprehensive Conservation and Management Plan for the San Francisco Estuary (Estuary Blueprint). https://www.sfestuary.org/wp-content/uploads/2017/08/CCMP-v26a-all-pages-web.pdf

SFBRWQCB. 2008. Appendix A: Revised Tentative Order. Bair Island Restoration Project, Redwood City, San Mateo County. California Regional Water Quality Control Board San Francisco Bay Region. https://www.waterboards.ca.gov/sanfranciscobay/board_info/agendas/2008/march/bair_island/bair_islandrevisedto.pdf

SFRWQCB. 2000. Beneficial Reuse of Dredged Materials: Sediment Screening and Testing Guidelines: Draft Staff Report. San Francisco Bay Regional Water Quality Control Board. https://www.waterboards.ca.gov/sanfranciscobay/water_issues/programs/dredging/Beneficial%20Reuse%20of%20Dredged%20Material_2019%20corrections.pdf

SFBRWQCB. 1998. Staff Report: Ambient Concentrations of Toxic Chemicals in San Francisco Bay Sediments. San Francisco Bay Regional Water Quality Control Board.

SFBRWQCB. 2019. San Francisco Bay Nutrients Watershed Permit, Revised Tentative Order No. R2-2019-0017. San Francisco Bay Regional Water Quality Control Board, Oakland, CA.

Smaal, A. C., & Nienhuis, P. H. 1992. The Eastern Scheldt (The Netherlands), from an estuary to a tidal bay: a review of responses at the ecosystem level. *Netherlands Journal of Sea Research*, *30*, 161-173.

SOTER (State of the Estuary Report). 2019. *Status and trends of indicators of ecosystem health*. San Francisco Estuary Partnership, Delta Stewardship Council. https://www.sfestuary.org/wp-content/uploads/2019/10/State-of-the-Estuary-Report-2019.pdf

SSFBS (South San Francisco Bay Shoreline Study). 2015. Final Integrated Document - Final Interim Feasibility Study with Environmental Impact Statement / Environmental Impact Report. Appendix G (Geotechnical Engineering). Prepared for U.S. Army Corps of Engineers by HDR Engineering Inc, Sacramento, CA.

Stern, M., Flint, L., Minear, J., Flint, A., & Wright, S. 2016. Characterizing changes in streamflow and sediment supply in the Sacramento River Basin, California, using hydrological simulation program—FORTRAN (HSPF). *Water*, *8*(10), 432.

Stern, M. A., Flint, L. E., Flint, A. L., Knowles, N., & Wright, S. A. 2020. The Future of Sediment Transport and Streamflow Under a Changing Climate and the Implications for Long☐Term Resilience of the San Francisco Bay☐Delta. *Water Resources Research*, *56*(9), e2019WR026245.

Stralberg, D., Brennan, M., Callaway, J. C., Wood, J. K., Schile, L. M., Jongsomjit, D., ... & Crooks, S. 2011. Evaluating tidal marsh sustainability in the face of sea-level rise: a hybrid modeling approach applied to San Francisco Bay. *PloS one*, *6*(11), e27388.

Sternberg, R. W., Cacchione, D. A., Drake, D. E., & Kranck, K. 1986. Suspended sediment transport in an estuarine tidal channel within San Francisco Bay, California. *Journal of Marine Geology, 71(3-4)*: 237-258.

Subtidal Goals. 2010. San Francisco Bay Subtidal Habitat Goals Report: Conservation Planning for the Submerged Areas of the Bay. California State Coastal Conservancy and Ocean Protection Council, NOAA National Marine Fisheries Service and Restoration Center, San Francisco Bay Conservation and Development Commission, San Francisco Estuary Partnership. http://www.sfbaysubtidal.org/PDFS/Full%20Report.pdf

Swanson, K. M., Drexler, J. Z., Schoellhamer, D. H., Thorne, K. M., Casazza, M. L., Overton, C. T., … & Takekawa, J. Y. 2014. Wetland accretion rate model of ecosystem resilience (WARMER) and its application to habitat sustainability for endangered species in the San Francisco Estuary. *Estuaries and Coasts*, *37*(2), 476-492.

Swanson, K. M., Drexler, J. Z., Fuller, C. C., & Schoellhamer, D. H. 2015. Modeling tidal freshwater marsh sustainability in the Sacramento–San Joaquin Delta under a broad suite of potential future scenarios. *San Francisco Estuary and Watershed Science*, *13*(1).

TBI (The Bay Institute). 2013. *Analysis of the costs and benefits of using tidal marsh restoration as a sea level rise adaptation strategy in San Francisco Bay.* https://bayecotarium.org/wp-content/uploads/cost-and-benefits-of-marshes-.pdf

Temmerman, S., Govers, G., Wartel, S., & Meire, P. 2004. Modelling estuarine variations in tidal marsh sedimentation: response to changing sea level and suspended sediment concentrations. *Marine Geology*, *212*(1-4), 1-19.

Temmerman, S., Bouma, T. J., Govers, G., Wang, Z. B., De Vries, M. B., & Herman, P. M. J. 2005. Impact of vegetation on flow routing and sedimentation patterns: Three dimensional modeling for a tidal marsh. *Journal of Geophysical Research: Earth Surface*, *110*(F4).

Thorne, K. M., Freeman, C. M., Rosencranz, J. A., Ganju, N. K., & Guntenspergen, G. R. 2019. Thin-layer sediment addition to an existing salt marsh to combat sea-level rise and improve endangered species habitat in California, USA. *Ecological Engineering*, *136*, 197-208.

UCB (University of California Berkeley), Stanford University, and San Francisco Estuary Institute. 2020. Reverse Osmosis Concentrate Treatment Research Results and Context for San Francisco Bay. Technical Memo.

USFWS. 2013. Recovery Plan for Tidal Marsh Ecosystems of Northern and Central California. Volume I. https://www.fws.gov/sfbaydelta/documents/tidal_marsh_recovery_plan_v1.pdf

USFWS 2015. Field Notes Entry: Levee Breach Celebrates Wetland Restoration on San Francisco Bay https://www.fws.gov/FieldNotes/regmap.cfm?arskey=36799

USGS (U.S. Geological Survey). 2013. CoNED 2m Topobathymetric Digital Elevation Model (DEM) of San Francisco Bay. USGS Coastal National Elevation Database (CoNED) Applications Project.

Verhoeven, J. T. A., Soons, M. B., Janssen, R., & Omtzigt, N. 2008. An Operational Landscape Unit Approach for Identifying Key Landscape Connections in Wetland Restoration. *Journal of Applied Ecology 45:*1496-1503.

Volk, R., Müller, R., Reinhardt, J., & Schultmann, F. 2019. An Integrated Material Flows, Stakeholders and Policies Approach to Identify and Exploit Regional Resource Potentials. *Ecological economics*, *161*, 292-320.

Wang, R. Q., Stacey, M. T., Herdman, L. M. M., Barnard, P. L., & Erikson, L. 2018. The influence of sea level rise on the regional interdependence of coastal infrastructure. *Earth's Future*, *6*(5), 677-688.

Ward, R. 1978. Floods: A Geographical Perspective. Macmillan, New York.

Watson, E. B., & Byrne, R. 2009. Abundance and diversity of tidal marsh plants along the salinity gradient of the San Francisco Estuary: implications for global change ecology. *Plant Ecology*, *205*(1), 113.

WRMP. 2020. San Francisco Estuary Wetland Regional Monitoring Program Plan. Prepared by the WRMP Steering Committee. San Francisco Estuary Partnership: San Francisco, CA.

Yarnell, S. M., Petts, G. E., Schmidt, J. C., Whipple, A. A., Beller, E. E., Dahm, C. N., … & Viers, J. H. 2015. Functional flows in modified riverscapes: hydrographs, habitats and opportunities. *BioScience*, *65*(10), 963-972.

www.ingramcontent.com/pod-product-compliance
Lightning Source LLC
Chambersburg PA
CBHW050917210326
41597CB00003B/128